T0302352

AVIAN NUTRITION

(POULTRY, RATITE AND TAMED BIRDS)

The book Avian Nutrition (Poultry, Ratite and Tamed birds) is enlarged edition of poultry and Ratite Nutrition. The scope of book has been increased by incorporating relevant and important informations on some of the popular tamed birds reared as pet in cages. Among these Parrots and Mayana are companion species capable of learning and limited conversation with members of the house. Bulbul, partridges and pigeon are used for sports. Owners of these birds expect guidance for care and management from the experts of poultry science because so far there appears to be no school on the management of such avian species.

Saras crane is state bird of Uttar Pradesh and its Socio-cultural significance has been depicted by statue of Saras at Saras Chauraha of Bharatpur district of Rajasthan. Swan is commonly known as Hansa or Rajhansa and ride of Goddess Sarswati of mythological significance. Incorporation of these avian species in the revised edition is expected to enrich the knowledge of students and teachers of the Avian and Poultry Science.

Nityanand Pathak (N.N. Pathak/N. Pathak) retired from laboratories on 30th June 2003 but maintained link with library. In a span of about half century studies on various aspects of veterinary Sciences he has tried to translate his studies in publications. These publications include more than 350 research papers, 30 books and many literatures of summer schools, short courses and Symposia etc. He had been President. Animal Nutrition Association Izatnagar, Animal Nutrition society of India, Karnal, vice president, Indian society of Buffalo Development, Hissar, and Indian Society of Zoo and Wildlife Veterinarians, Bareilly. He was also General Secretary Animal Nutrition Association, Izatnagar and National Academy of veterinary Science, New Delhi, At present he is president, Pashu Poshan Kalyan Samiti and Academy of veterinary Nutrition Izatnagar. He is the recipient of Prize of Commonwealth Bureau of Animal Hearth, England; Jawahar Lal Nehru Award of ICAR, New Delhi; University level Best Teacher award, IVRI, Izatnagar; Invention Award NRDC, New Delhi and WIPO award, Geneva.

AVIAN NUTRITION

(POULTRY, RATITE AND TAMED BIRDS)

By

NITYANAND PATHAK
B.Sc., B.V.Sc. & A.H., M.V.Sc., Ph.D.,
FNVS, FANA, FMob S,
113 B Vatika Suncity Vistar,
Izatnagar, Bareilly-243122 (UP)

NARENDRA PUBLISHING HOUSE
DELHI (INDIA)

First published 2021
by CRC Press
2 Park Square, Milton Park, Abingdon, Oxon, OX14 4RN

and by CRC Press
6000 Broken Sound Parkway NW, Suite 300, Boca Raton, FL 33487-2742

CRC Press is an imprint of Informa UK Limited

British Library Cataloguing-in-Publication Data
A catalogue record for this book is available from the British Library

Library of Congress Cataloging-in-Publication Data
A catalog record has been requested

ISBN: 978-0-367-69447-0 (hbk)
ISBN: 978-1-003-14184-6 (ebk)

Contents

Preface

Chicken production for nutritious table eggs and low fat meat is now a well established commercial enterprise occupying fifth and sixth positions in egg and meat production in the world. Indeed production is much less than the requirements. Due to increasing prices of foods it is becoming difficult for a considerable population to include proteinous foods of high nutritive value. Revival and extension of backyard poultry farming on scavenging with partial supplemental feeding may be helpful in increasing nutritional levels of such families. Introduction of large size ratite avian species may be economical due to capacity of these birds for extracting nutrients from the herbaceous forages. The enteric symbiotic microbial inhabitants provide ability of digestion of fibrous (cellulosic) feeds.

Basic information's on social birds like Parrot, Pigeon, Partridges, Mayana, Peo fowl, Bulbul, Saras crane and Swan. will enrich the academic capacity of the natural for a family having a parrot in cage to know more about the foods, habits and behavior of the bird. Such aspects have been ignored in most of the titles on poultry and other avian species.

In the preparation of this book assistance and comfort provided by Anuvrat and Priyabhashni helped to complete the task in short time. Dr. S.K. Saha helped in collection of some information's from the library. The assistance from different corners is acknowledge.

Nityanand Pathak

1 INTRODUCTION

Poultry production for nutritional security of mankind has been established as one of the quick method of nutritious food production of high biological value. The coordinated application of breeding, feeding, management and maintenance of health has made possible to finish broilers of 1 to 2 kg body weight in 4-6 weeks and many layer strains of fowl are now laying more than 300 eggs annually. For sustaining such high level of productivity the requirement of specialized education and intensive training has acquired important place in the syllabi of veterinary science and animal husbandry. Feeding alone costs about 70 per cent of total expenditure of commercial poultry production. Therefore, it is important to understand the feed utilization processes in fowl and other birds.

Majority of Indian population was strict vegetarian and milk, milk products and pulses were the sources of dietary protein. However, this pattern has changed at a fast rate during the past half century and even strict vegetarian families are now not averse of inclusion of eggs and even many other kinds of foods of animal origin in the diets of youngsters in the family. The human population has crossed 121 million number by the end of 2010. It is becoming difficult to bring down population growth below 18 per thousand. At the same time average age is also increasing. Under such situation requirement of highly nutritious foods of animal origin will continue to increase for maintaining the requirement of growing children, pregnant ladies and senior citizens.

Factors responsible for increase in the number of non-vegetarian population

Some of the factors found responsible for increase in the number of non-vegetarian population is due to the major shift in the food habit of vegetarian population which may be listed as follows.

1. **Drastic fall in the production of pulses:** The pulses were an important and easily available source of dietary protein in the diets of almost all categories of vegetarian and other people because pulses were in the category of coarse grains and they were much cheaper than most of the cereal grains, but the balance changed and now pulses are beyond the reach of common people. No doubt that during 1960s there was acute need of increasing the production or cereal grains to save the people from starvation. The success of green revolution was praised globally but the extension of approach for very long period disturbed the balance of food grain production. Monoculture that too of wheat and rice for food, and sugarcane, toacco and cotton for commercial farming changed the composition of human kitchen. This resulted in craving for proteinous foods and ultimately gradual shift to protenious foods of animal origin started and today good number of younger members of even from strict vegetarian background are eating the preparations of egg, meat and fish. Such members eat such foods either in restaurants or in some houses separate kitchen and utensils are used.

   The composition of food crops has been seriously disturbed due to delay in the analysis of food grain production pattern. The corrective measures for balancing the ratio of cereal grain, pulses, oil seeds and commercial crops are considerably delayed and it will take some time to place the situation in order.

2. **Increasing population and changed demography:** Under the present programmes of family welfare there appears to be little scope of considerable fall in the growth rate of human population. The present growth rate of 17 or 18 births per 1000 is actually would have been a little higher, if there could have been effective control on the female foeticide. The increase in proportion of infants nursing mother, growing children and geriatrics will proportionately increase the requirement of proteinous foods of animal sources specially the milk and egg.

3. **Exposer to nutritional significance of foods of animal origin:** Due to increasing education, interaction among population, different extension agencies, family welfare programmes and media the awareness about the nutrition and significance of foods of animal sources in the diets of humans is increasing.

4. **Shrinking of arable land:** Due to increased diversion of agricultural land for infra-structure development, urbanization and change in life style a considerable area of land is being used for industries, housing, markets, educational institutions and health care centres etc. Due to division of families into nuclear family the space in house like common room or so called drawing

room and kitchen etc is multiplying. The impact of such social changes will be felt seriously if carrying capacity of land for food production and animal production (including poultry) is not increased with the comparable pace.

5. **Busy life:** the time is becoming a limiting factor in house management specially in urban life requiring all eligible member to earn for running the family. Morning is hectic in almost all households and it has been seen that majority of urban house holds are using boiled eggs, omlets, bread, butter or cheese and some fruits for breakfast and also for the lunch of school going children. Due to lack of time in morning it is very difficult for the working couple to serve conventional foods. This change in life style has significantly increased the demand of eggs specialy in the urban areas. The families also feel shy in offering traditional fast foods of local menu due to fear of being ticked as backward or miser.

6. **Socio-economic change:** There has been fast change in the economic status of a new service class persons employed in national and multinational companies mostly situated in big cosmopolitan cities. The persons employed in such organizations are mostly on high pay-package but hard press with work load besides requiring long time for attending office or work place due to congestion and traffic problems. The expenses on house hold management, children education and maintenance of health are increasing uncontrolled and it has become difficult for a so called middle class group to meet the normal family requirements which necessitate earning by all eligible members in the family. In present set up of cosmopolitan cities it is mostly young and middle age couple. This group is mostly dependent on fast foods or easy to prepare in short time. Such situation has increased the consumption of eggs.

7. **Changed life style:** In recent years it has been seen that youngsters are joining jobs at younger age, remain unmarried for longer period and stay away from the parents at the place of job. This group of people are now mostly dependent on hotels, modern dhabas and restaurants. In such groups consumption of non-vegetarian foods and soft to hard drinks has increased beyond imagination. This group is so much engaged that there is no time for considering the pros and cons of such life and knowing the quality of food consumed.

Why poultry production

The high nutritive value of eggs and chicken, fast turn over of production and easy availability are the factors favouring the growth of poultry production, viz.

1. Broilers can be harvested at 4, 5 or 6 weeks of age depending on the market demand.
2. A well bred hen starts laying at about 22 weeks of age and continue to lay efficiently for 15-18 months.
3. At the end of laying age the spent hens are used for table purpose. The palatability and digestibility improvement has been made possible with the use of tenderizing herbs, fruits and enzymes etc.
4. Preparations of eggs are many and easy to prepare in short time.
5. Spent hens are used for the preparation of value added foods of longer self life.
6. Marketing is not a limitation. Even in remote villages one can get eggs and routine egg products like boiled egg, poached egg and omlet etc.
7. Low capital requirement for establishing small poultry farming units for livelihood earning.
8. Low capital requirement for marketing of egg preparations for livelihood earning.
9. Increased scope of the application of newer technologies for the production of designer eggs of required composition.
10. Backyard poultry helps in providing fresh eggs for domestic consumption or subsidiary income in rural areas.
11. Poultry manure is nutritious for crops in rural areas and pot plants in the urban areas.

Different species of poultry birds

Fowl is the main poultry bird distributed world over for the production of table eggs and chicken meat. Now numerous strains of white leghorn fowl have been developed for higher egg production and better feed conversion efficiency. For excellent meat production composite broiler strains have been synthesized from the cross breeding of highly prolific white leghorn and heavy meat producing breeds. In addition to chicken several other species of birds have been developed as poultry birds.

The term poultry is applied to the group of domesticated avian species reared for the production of edible eggs and meat. The different species of poultry birds may be classified into the following two major groups on the basis of habitat.

1. **Terrestrial poultry birds:** The avian species living in dry environment are the terrestrial poultry birds. These may be further classified into the following two sub groups.

   (i) Common terrestrial poultry bird species. These are chicken or fowl, guinea fowl, quail and turkey.

   (ii) Not yet fully recognized poultry bird species are pigeon, dove, pheasants etc.

2. **Semi–aquatic poultry birds:** The habitat of this group is the wet land along the water sources. These birds like to swim and dive in the water and during this game they feed on small fishes, lobster, snails, tadpole, spawn and many other aquatic animals. Many species of aquatic birds eat significant amount of aquatic herbages like lemna, azolla, hydrilla, chlorella, spirulina and many others. Duck is the most common aquatic poultry bird which has been developed to lay about 200 eggs annually. The other domesticated aquatic poultry birds are the goose and swan. Duck rearing is more popular in the paddy growing tropical countries where this species is fattened on the fallen grains following paddy harvesting. During scavenging in paddy fields the ducks get different insects, small snails and other sources of animal protein and minerals.

   Duck rearing is preferred over goose and swan rearing because the later become often aggressive with small children, production is low and feed conversion efficiency is also low. The duck is docile, smaller and more efficient feed converter.

Important native (desi) chicken or fowl in India

Inspite of tremendous qualitative and quantitative development in chicken production in India the craze for eggs of deshi hen and meat of deshi cock widely persists specially in the rural areas extending from Gujarat in west to north-east hill states in the east. The eggs and meat of native chicken are also popular in the neighbouring countries like Nepal, Bangladesh, Bhutan and Mayanmar.

Both egg and meat of native chicken, the domesticated descendents of Indian red jungle fowl are liked. It is a common belief that the eggs and meat of native birds reared on scavenging are more nutritious and tasty. Both these products fetch considerable higher price when compared to eggs and broilers produced on farms from the developed high producing lines of chicken. The demand of eggs of deshi fowl has induced mal practice of faking eggs.

Faking of eggs for table purpose

The egg production of native strains of Indian fowl is very low and range from 40 to 80 eggs per year in the traditional flocks of different areas reared in backyard management. The eggs of Indigenous Indian chicken are small (40-50g), more round in shape with brown colour shell. The local retail dealers short out smaller size brown shell eggs of poultry (fowl) like Rhode Island Red, Australorp, Cornish, New Hampshire and Plymouth rock etc. For larger size consumers are convinced for better feeding and management due to awakening of the small holder poultry farmers.

The another method of faking egg is the colouring of white shelled eggs. Smaller and relatively more round eggs of farm hen are shorted out, cleaned and then dipped for 2-3 hours in tea leaf extract to become brown. Light extract of coffee is also used for the purpose. Since tea and coffee are now very costly items the faking method has changed to painting by dipping in solution of synthetic brown paint. However, it is not easy to bluff rural people but also it is not difficult to cheat majority of the urban consumers not exposed to shape, size and colour of eggs of native fowl.

The third method is not popular but many times used in the urban markets of coastal area to mix the eggs of turtle and tortoise which are easily found covered beneath sand in coastal area during the laying season.

Important by-products of economic value

Besides nutritious foods of high biological value different useful by-products of economic importance are also available from poultry farming. Most of the by-products contribute little in the earning of very small farmers keeping few birds in back yard system but contribution of by products becomes significant in organized farming. The various by-products of agricultural and industrial importance are recorded as follows:

1. **Cage bird droppings:** Poultry droppings collected from caged birds is a rich source of nitrogen and minerals. The nitrogen content ranges from 3 to 4 per cent in dry matter of manure from different class of birds being higher in the dropping of starter and grower chicks of layers and the broilers. Calcium content is very high in the droppings of laying birds. The content of phosphorous, potassium, magnesium and trace minerals zinc, copper, manganese, iron and iodine etc. is also high. Due to high concentration of dietary essential minerals the sterilized germ free dried and powered droppings

of caged birds are also used for the manufacture of mineral mixture for the ruminants because the non protein nitrogenous component in the droppings can be utilized by ruminants only for the synthesis and supply of proteins.

2. **Poultry litter:** The raising of poultry birds on litter is still used by the small and medium farmers. The bedding material used are chaffed straw or dry grasses, wood savings, dry fallen tree leaves and saw dust. The bedding material is enriched by the droppings of birds. Litter is normally turned at an interval of 3-4 weeks and replaced after about 3 months. This litter is composted and used for the manuring of grain and vegetable crops.

3. **Slaughter house by-products:** These are feather, legs and some of the internal organs like respiratory tract and gizzard removed digestive tract. These offals are offered fresh in the feed of swine but at large slaughter house these are processed for the preparation of protein and mineral supplements of livestock. Feather is usually separated for other commercial uses.

4. **Feather:** The feathers of poultry collected at slaughter house are shorted in different classes on the basis of size, softness and colour. These are used in cottage industry for the preparation of decoration pieces, fan, feather plume for cap etc. Small size soft feathers are used for the stuffing of cushion, pillow and mattress. After sorting, the bulk of feathers is also hydrolyzed for the manufacture of feather meal as a protein supplement in the diets of pig and other simple stomached animals.

5. **Hatchery residues:** These include empty shells, unhatched eggs and dead embryos. These are sterilized, dried, ground and standardized for marketing as a hatchery waste meal for the feeding of simple stomached animals and birds. Hatchery waste meal is a rich source of calcium and other minerals, animal protein and some of the vitamins. The products are variable in chemical composition and it should be shown on the packing. At large slaughter houses non edible offal available from the slaughter of poultry birds are processed and then standardized for chemical composition before marketing. Unprocessed offals available from the slaughter of few birds are fed either raw or after heat treatment (cooking). Feeding of raw offal and offals stored in open is highly risky and may be cause of dreaded disease.

Nutritional constraints in poultry production

Medium and large poultry farms are run by economically well placed houses capable of managing necessary inputs including balanced compounded feeds and supplements. Small poultry farmers and particular those rearing small flocks in backyard management in remote areas frequently fail to provide adequate feed of optimum feeding value. This results in slow growth, delayed start of laying, smaller eggs, shell less eggs and less number of egg production. The effect of confined management on reduction in production is normally more in comparison to scavanging birds having greater opportunity for picking insects, larvae, small snails and other stray sources of animal protein and minerals. However, excess use of insecticides and pesticides and uncontrolled disposal of hospital wastes is making the system of scavenging dangerous not only for the poultry birds and wild birds but also for the owners of such birds and consumers using the eggs and meat of such birds.

Impact of indiscriminate use of chemicals and drugs and uncontrolled disposal of hospital wastes

The use of insecticides and pesticides is not only increasing but their concentration is also increasing. This is polluting the environment and natural resources. The other major sources of harmful drugs and chemicals are the hospital wastes. There is no strict monitoring of the proper aseptic disposal of hospital wastes. The next source of contamination is the disposal of carcasses of dead animals in open and even near the sources of water. These are responsible for the vanishing of birds like vulture, kite, owl and accidental killing of scavenging field birds like sparrow, parrot, pigeon, dove and peacock etc. Whenever there is strictness on the proper disposal of hospital wastes, the place of disposal is mostly shifted to a remote place of rural area away from the common roads. Under such situation it is always safe to keep the birds in confinement in an enclosure at home or inside the poultry house. This practice will undoubtedly increase the expenditure on feeding and lower earning but definitely provide protection of bird's health and also the health of consumers utilizing eggs and meat of such birds. Now a days it is also risky to hunt and consume game birds perching on shrubs and trees in the vicinity of cropping fields.

Birds for food

Mankind was dependent on herbages and animals for food since its evolution. The selection of natural food sources started with the start of civilization. Birds are one of the main food source of humans all over the world. Both wild and domesticated birds are used for food. Hunting of different species of birds was popular through out the world and for many species extensive hunting of game birds threatened their existence. This situation gave birth to wild life protection movement in many countries and ultimately many societies are now working in different countries to protect the threatened species.

Game birds

The wild avian species hunted for food are known as game birds.

Domesticated birds

The avian species now bred and maintained by humans for different purposes are called domesticated birds. Domesticated birds are placed in different categories according to their uses such as poultry for food (Food bird), pet (Caged) birds, ornamental birds and show (zoo) birds etc.

Classes of birds according to habitat

The free living birds are also placed in the following groups on the basis of the habitat used in natural conditions by the species. These are

1. **Arborial birds:** The birds making home for living and multiplying on the trees and also those resting on the trees in night are known as arborial birds. This is one of the largest group.
2. **Terrestrial birds:** The members of this group are seen feeding on the fallen grains, insects and succulent herbages in the fields. These are mostly nesting birds but also include birds living in abandoned buildings, temples, churches, mosques etc. Some small birds live in small holes in caves and some other make holes in mounds of soil.
3. **Aquatic birds:** This is also a large group of avian species. The birds spent greater part of the day swimming and diving in the water bodies. These include ducks, geese, swan and many others. These birds lay eggs in bushes along the water sources and most of the species also retire at safe places along the water bodies or in the bushes. A few species of aquatic birds take shelter on trees during night.
4. **Arcatic birds:** These are cold blooded birds capable of adjusting body temperature according to environmental changes in the arctic region. Penguin is the typical example which lives standing about 6 months at a place during the cold season.

5. **Migratory birds:** These are avian species of cold zone capable of migrating long distance intercontinently. Migration of temperate birds to tropical country is an annual practice. These are mostly aquatic birds that migrate to tropical region during the winter season. Siberian cranes and other birds of northern Asia and Europe visit every year and flock on the water sources particularly the fresh water ponds and lakes. Breeding and hatching also occur in the migratory birds. These birds have been found to have well developed sense organs and they visit the same place every year. Change of place occurs only in unfavourable conditions.
6. **Ratites:** These are almost non-flying birds of generally large size in which keel bone is not formed, e.g. Emu, Ostrich, Rhea and Kiwi.

Scientific or zoological names of some birds

In order to avoid confusion in identification of birds including poultry birds it is important to use scientific (zoological) names which do not change. Binomial and trinomial system of nomenclature are used for the names of plants and animals. Scientific names of some common poultry and few other bids are given in Table 1.1.

Table 1.1: Zoological names of some birds

Common name	Zoological name
Wild fowl	Gallus gallus
Domesticated fowl (Chicken)	Gallus domesticus or Gallus gallus domesticus
Duck	Anas platyrhynchos
Muscoy duck	Cairina moschata
Guinea fowl	Numida meleagris
Goose	Anser anser
Swan	Cygnus spp.
Quial	Coturnix spp.
Turkey	Melleagris gallopavo
Partridge	More than 22 genera and many species
Parrot	Psittacus erithacus
Pea fowl	Pavo cristatus
Dove	Columba oenas
Pigeon	Columba livia
Pheasants	Phasianus colchicum
Emu	Dromaius novachollandiae
Ostrich	Struthio camelus
Rhea	Rhea americana and Rhea pennata
Kiwi	Apteryx spp.

Ancestors of some domesticated and tamed birds

The ancestors of common fowl or chicken (Gallus domesticus) are more than one and they are still found in the jungles of Indian sub continent, east Asian and south east Asian countries. The most important are the red jungle fowl and Brahma fowl of India. Some informations about the native land and purposes of domestication are presented in Table 1.2.

Table 1.2: Ancestors and purpose of domestication of some avian species.

Domestic bird	Wild ancestor (s)	Country of domestication	Purpose of domestication
Common fowl (Chicken)	Red junble fowl Brahama fowl, Grey jungle fowl, Naked neck fowl, Miri fowl Sonnerat's fowl	South Asian and south east Asian region	Eggs, Flesh, feathers, bird fight, manure
Duck	Mallard duck and Muscovy duck	Many coutries	Meat, eggs, feather
Goose	Greatleg goose and swan gooe	Do	Meat, eggs, feather, ornamental
Swan	Mute swan	Do	Meat, eggs, feather, ornamental
Domesticated Guinea fowl	Helmeted guinea fowl	Africa	Flesh, eggs, feather and alarm calling
Turkey	Wild turkey	Mexico	Meat, feather
Emu	Wild Emu	Australian continent	Meat, oil, feather, egg
Ostrich	Wild Ostrich	Australian continent Africa	Meat, eggs, feather, leather, ride
Pea fowl	Wild peafowl	Indian sub continent	Landscaping, Ornamental
Pigeon	Wild pigeons	Many countries	Joy, messenger, ornamental, meat
Partridge	Wild partridge	India, Many countries	Bird fight show, pet, meat
Parrots	Wild parrots	Many countries	Caged pet, ornamental, meat
Pheasants	Common pheasonts golden pheasant cheer pheasant	Many countries	Ornamental, meat
Mayana or Myanah	Common mayana Himalayan Mayane	Asia (India etc)	Caged pet for talking, ornamental
Bulbul	Red vent yellow vent	Indian sub continent	Caged pet for bird fight show

Breeds of poultry birds

Extensive research works have been done in different countries to develop chicken, other poultry birds and also some other avian species as food birds. Poultry farming is now one of the important profession of the world. The high prolificacy, quick turn over and small size have made poultry an important source for solving the protein problems of the human dietary. A large number of original breeds, crossbreeds and strains (varieties) capable of producing large number of eggs (300 or more per year) and fast growing broilers (broiler chicken 1.5 kg or more in 6 weeks). It is in interest of the poultry professions to know the breeds of birds which have been compiled in Table 1.3.

Table 1.3: Breeds of poultry and some other food birds

Bird (Species)	Breeds/ Varieties
1. Chicken/ Fowl (Gallus domesticus)	(A) **Light breeds:** Anona, Andalusion, Barred rock, Leghorns (Varieties are White Leghorn, Black Leghorn, Brown Leghorn, Exchequer Leghorn), Minorca (Black Minorca, White Minorca), Legbar (Brown leghorn x Barred Rock) and Welbar (Welsummer x Barred Rock).
	(B) **Light Exhibition breeds:** Canpine, Coveney, Derbyshire, Hamburg, Labresse, March Diasies, Polish, Scots, Sicilion, Spanish and yak.
	(C) **Heavy utility breeds:** Astralarp, Bamvelder, Orpington (Black, Blue, Buff and White), Plymouth Rock (Barred, Black, Buff and White), Rhode Island Rd (also Black and White), Sussex (Light, Speckled, Red, Brown, Buff), Welsummer and Wyandotte (many varieities).
	(D) **Heavy Meat breeds:** Dorking (Single comb, Rose Comb, Cuckoo, Dark Red, Silver gray and White (Toes number 5), Faverolle (Blue, Buff, Salmon and White), Old English game, modern English game and Maran.
	(E) **Exhibition Heavy breeds:** Brahma (several varieties), Cochin (Black, Buff, Cuckoo, Partridge and White), Indian game, Jersey Black giants, Langshan (Groad, Modern Black, Modern Blue and Modern White), Maline (Maline Proper, Turkey headed Maline), Silkie (Black, Blude, pure white).
	(F) **Indian breeds:** Aseel, Black Bengal, Brahma, Busara, Chitagong, Delhem Red, Deniki, Ghagus, GiriRaja, Gram Lakshmi, Grampriya, Kadaknath, Kalahasti, Kalinga Brown, Kahmine Feverella, Mumbai desi, Naatikori (Kudla), Naked neck, Nicobari, Tellichery, Titasi and Vanaraja.

[Table Contd.

Contd. Table]

Bird (Species)	Breeds/ Varieties
2. Duck	Abacot Ranger (or Streicher), Allier or (Blanc d' Allier), Ancona, Aylesbury, Baliduck, Black East Indian, Blue Swedish, Buff Orpington, Black Orpington, Blue Orpington, Call duck, Cayuga duck, Challans duck, Chara chembali, Creasted duck, Danish duck, Duclair, Dutch Hookbill, East Indian duck, Forest duck (Eend Van vorst), Gimbsheimer, Golden cascade, Greyssingham (Wild Makard x Pekin), Huttengem, Indian and Pencilled), Jaime Ledesma, Khaki Compbell (also white Campbell), Magpie (Black and White varieties), Major can, Muscovy, Pekin or (Long Island), Rouen, Saxony, Semois, Silver Appleyard, Silver bantam, Termonde, Venetian (Germanata veneta), Welsh Harlequin, Wood duck, Swedish duck and Swedish yellow duck.
3. Goose (Geese)	African goose, Bar-headed goose, Brecon buff, Chinese goose, Emden, English, Grey Chinese, Roman, Sebastopot, Toulouse and White Chinese.
4. Swan (Species)	Black naked swan, Black swan, Bewick's swan, Indian Swan, New Zealand swan, Trumpeter swan, Tundra swan and Whooper swan
5. Guinea fowl	Lavender, Pearl and White
6. Quail	Black breasted, Blue breasted, Californian, Chinese painted, Common quail, Common Bustard gray, Hereliquin, Japanese, Jungle, Little Bustard, Manipur Bush, Mountain, New Guinea mountain, New Zealand, Painted bush Rain, Rock bush, Yellow leg button.
7. Turkey	Broad breasted white, Broad Breasted Browgze, Bourbon Red, Beltsville small white, Blue slate (or slate), Chocolate, Midgate white, Narragansett, Norfolk black, Spanish black and Standard bronze.
8. Pigeon	Archangel (Or gimpel), Carriers, Damascine (or Mahomet), Drummer (Or Trumpeter), Dutch Capichin, English Swift, Foot Soldier (Or Flying Serpent), Helmet, Indian Fan tail (or Fan tail), Jacobin, Komorner Tumbler, Modena, Nun and Old German owl pigeon.
9. Partridge	Twenty two genera are Alectoris, Ammoperduix, Anurophasis, Aborophila, Bambusicola, Caloperdix, Coturnix, Francolinus, Glloperdix, Haematortyx, Lerwa, Margaroperdisc, Ophrysia, Perdicula, Perdix, Ptilopachus, Rhizothera, Rollulus, Tetraogallus,l Tetraophasis and Xenoperdix. There are large varieties/ breeds.

[Table Contd.

Contd. Table]

Bird (Species)	Breeds/ Varieties
10. Pheasnants	Many
11. Emu	Four subspecies. Australia
12. Ostrich	North African, South African, Massai, Arabia and Somali
13. Rhea	Rhea americana and Rhea pennata
14. Kiwi	Brown, Great spotted, Light spotted, Okarita and Montelli brown. New Zealand
15. Bulbul	Red-Vented and Yellow-vented

Terminologies for different stages etc.

Different terms are used for the differentiation of male, female, chicks, growers and castrates. There are also specific terminologies for some conditions of the birds. Important terminologies are presented in Table 1.4.

Table 1.4: Terminologies used for stages of poultry.

Species	Male	Female	Young	Castrate
Fowl	Cock	Hen	Chick	Capoun
Pea fowl	Peacock	Peahen	Pea chick	
Guinea fowl	Male guinea fowl	Female guineafowl	Keet	
Duck	Drake	Duck	Duckling	
Goose	Gander	Goose	Gosling	
Swan	Cob	Pen	Cygnet	
Turkey	Tom turkey	Hen Turkey	Poult	
Growing fowl	Cockrel	Pullet	-	
Quail	Quail cock	Quail hen	Quail chick	
Pigeon	Male Pigeon	Female Pigeon	Squab	
Emu				
Ostrich				
Rhea				

Some important characteristics of modern chicken (Gallus domesticus) breeds

Some important phenotypic traits are used for the identification of breeds. Apparent differences are found in the size and colour of different body parts among the breeds and strains of chicken. Almost all breeds of American chicken class lay brown shelled eggs. In English class of chicken egg shells may be brown or tinted. White shelled eggs are laid by Mediterranean breeds and small size brown shelled eggs are laid by wild as well as domesticated Asiatic and Indian breeds (Table 1.5).

Table 1.5: Physical traits of some modern chicken breeds

Breed	Comb type	Colour of some body parts				
		Ear/lobe or wattle	Skin	Shank	Egg shell colour	Feathering of shank
A. American chicken breeds						
Jersey black giant	Single	Grey red	Yellow black	Brown	Brown	Absent
New Hampshire	Single	Red	Yellow	Yellow	Brown	Absent
Plymouth rock	Single	Red	Yellow	Yellow	Brown	Absent
Rhode Island Red	Single/Rose	Red	Yellow	Yellow	Brown	Absent
Wyandotte	Rose	Red	Yellow	Yellow	Brown	Absent
B. English Chicken breeds						
Astralarp	Single	Red	White/ Black	Black	Brown	Absent
Cornish	Pea	Red	Yellow	Yellow	Brown	Absent
Dorking	Single	Red	White	White	Tinted	Absent
Orpington	Single	Red	White	Bluish	Brown	Absent
Sussex	Single	Red	White	White	Tinted	Absent
C. Mediterranean breeds						
Ancona	Single	White	Yellow	Yellow	White	Absent
Andalusian	Single	White	White	Black	White	Absent
Leghorn	Single	White	Yellow	Yellow	White	Absent
Minorca	Single	White	Black	Black	White	Absent
D. Asiatic/ Indian breeds						
Brahma	Pea	Red	Yellow	Yellow	Brown	Present
Cochin	Single	Red	Brown	Blackish	Brown	Present
Langshan	Single	Red	Brown	Yellow	Brown	Present

2

FEEDS OF POULTRY

Feeds are normally the natural organic sources eaten for the supply of nutrients essential for the maintenance of optimum physiological functions of the body.

Main sources of poultry feeds

Poultry birds may be omnivorous and herbivorous. Fowl (Chicken) is the main poultry bird around which commercial farming is flourishing in the country. The other important species of the Indian poultry industry are the duck and quial. Guinea fowl, turkey and recently introduced emu are yet to make a place in production. The later species are mostly grazer and consume considerable quantity of soft and leafy herbages. Thus, the sources of feeds for poultry and ratites may be enumerated in the following groups.

1. Feeds of plant origin
   - (a) Food grains, pulse, oil seeds and their milling by products.
   - (b) Succulent herbages
   - (c) By-products of sugar industry
   - (d) Aquatic plants
   - (e) Fruits and vegetable shortings
2. Feeds of animal origin
   - (a) Meat and offals of mammalian food animals
   - (b) Meat and shortings of aquatic animals like fish, crab, lobster, frog and snail etc.
   - (c) By-products of dairy industry
   - (d) Poultry eggs and carcass processing by products
   - (e) Silk worm pupae meal

The compounded commercial feeds are normally complete balanced diets manufactured for the feeding of different classes of poultry for egg and meat production. In the preparation of compounded feeds the nutritional requirements for different production functions are the main consideration. Therefore, feeds are classified as follows:

1. ENERGY RICH FEEDS

Yellow maize is considered the main feed source around which poultry industry develops. But most of the other food grains are comparable sources of energy. The energy feeds may be further classified into the following sub groups.

A. **Cereal grains:** Maize, wheat, rice, sorghum, milo, pearl millet, barley, oats, finger millet, paspalum and other minor millets.

B. **Cereal grain milling by-products:** Wheat bran, rice bran, rice polish, grain shortings, maize gluten, maize gluten feed, broken rice etc.

C. **Sugar:** Jaggery and molasses.

D. **Root crops and tubers:** Beat root, turnip, tapioca, yam, potato, sweet potato, carrot

E. Seeds of fruits and forest like sal seed meal, mahua seed cake, mango seed kernel cake, tamarind seed, nahor seed cake, azar seed cake, karanj cake, castor bean cake, neem seed kernel cake etc.

F. **Fats and oils:** These are richest sources of dietary energy and used for balancing the dietary energy level of the diets of high laying hens and fast growing broilers.

Cereal grains

Cereal grains are the major constituents of the compounded commercial feeds of simple stomached animals like pigs and poultry birds. Some of the common cereal grains used for the feeding of poultry are maize, wheat, rice, sorghum, pearl millet, pearled barley, dehusked oat, finger millet (Ragi or Madua), Paspalam and other minor millets. Yellow maize is the most preferred food for the feeding of laying hens because the egg yolk of such hens are golden yellow and fetch higher price. Wheat, rice, sorghum and other grains are used in the diets of broilers and other birds raised for meat production.

Maize or corn (Zea mays)

Poultry farming develops around maize grain. The common colours of maize grain are white, yellow and reddish. There are hard and soft as well as sweet varieties.

Yellow maize is preferred for the feeding of laying hens for obtaining eggs of golden yellow or yellowish orange colour yolk for table purpose. Maize contains precursors of vitamin A, cryptoxanthine and carotenes. Oil content ranges from 3 to 5 per cent (average 4%) and contains high proportion of linoleic acid which is required for the maintenance of size and production of eggs. Crude fibre is nominal (about 2%) and starch content is quite high (71-75%) due to which ME value is highest (3350 kcal per kg) amongst the cereal grains. The protein content is highly variable and ranges from 9 to 14 percent in dry matter. The quality of protein (except opaque 2 variety) is also poor. The two major types of maize protein are zein and gluten. The former is a constituent of endosperm and deficient in some essential amino acids like lysine and tryptophane. The other protein, maize gluten is mostly found in embryo and small amount in the endosperm.

The superior variety of maize opaque 2 is a richer source of lysine and tryptophane and its advantage occurs only on methionine supplementation.

Another newer variety floury -2 contains higher quantity of both, lysine and methionine.

Harvesting management and storage of maize grain is very important because it is highly susceptible for fungal infestation. The common fungi infestation is that of Aspergillus flavus and A. parasiticus which produce highly detrimental aflatoxins. The other may be Fusarium sp. producing zearalenone. A third group is the species of Fusarium that produce ochratoxins. Therefore, it is very important to procure healthy maize grains containing less than 11 per cent moisture. The store should be well aerated, dry and free from insects and rodents.

In some countries maize is not fed to broilers and other birds reared for meat production because fat becomes yellowish and greesy or loose and lowers the carcass quality.

Sorghum or jowar (Sorghum bicolor/S. vulgare)

This is an important cereal grain crop of rainfed area of low annual precipitation. The two varieties cultivated widely in India are the white and light yellow or pale yellow.

The other varieties of brown and red colour (milo) contain higher level of tannins and palatability is lower but these are bird resistant. White and yellow sorghum is used to replace significant amount of maize from the poultry feeds. Protein content is higher, oil content is less and fibre and tannin contents are variable. Higher tannin content reduces palatability. Wild birds pick up seed of red and brown variety only when other feeds are not available. Average ME values of white, yellow and red sorghum are 3050, 3000 and 2700 kcal per kg dry matter respectively. Some difference in energy utilization may be seen among different poultry species. Tannins content ranges from 0.2 to 3/% or even more, crude protein 10 to 13% and crude fibre 2 to 2.5%. Calcium content 0.5% is much higher than 0.3 % in maize. Almost all essential amino acids are less in comparison to maize.

Self life of dried wholesome grain containing less than 11 per cent moisture is much higher than the maize under identical storage. Whole grain is fed to caged ornamental birds and grit feeding can be ignored without adverse effect on health. Attack of wild parrots and other grain eating birds is quite high on the sorghum crop specially at dough to mature stage.

Pearl millet or bajra (Pennisetum typhoides)

It is a sturdy rainfed crop of semiarid region. Pearl millet is staple food for humans in some parts and also used for energy supplementation in the diets of livestock and poultry. Whole grain is fed to backyard poultry and also to ornamental birds. The grains are brown rough coated.

Crude protein content of bajra is higher than the common maize and jowar. It is fed as such to small flocks of poultry maintained in backyard management and foraging. Bajra is rarely used for energy supply to replace maize from the commercial compounded poultry feeds. However, small scale feed manufacturing companies use pearl millet upto 30 percent for the substitution of maize from the layers feeds.

Rice and paddy (Oryza sativa)

Rice is a rich source of carbohydrates and oil. Coarse rice with polish prepared by indigenous milling methods is fed to poultry alongwith other feed ingredients. Broken rice is used to replace varying proportion of maize from the compounded diets of poultry. Proximate composition is variable and crude protein content ranges from 7 to 10 percent. ME value of rice is comparable with maize, wheat and white jowar. It is used to replace 25 to 50% maize from the poultry feeds when available at a competible price.

Foraging birds like duck, geese and swan pick up fallen paddy grains and insects. It is a common practice to run the flock of growing and adult laying duck flocks into the paddy fields following harvesting. It has two major benefits. The fallen whole grains are efficiently picked up. The insects, snails and larvae are the sources of animal protein and minerals. The second benefit is the dispersal of nutritious organic manure of duck in the field. Foraging is followed by ploughing and transplantation of paddy seedlings. This is a continuous process in the hot humid climate of south-east Asian countries. There is no problem of oxidative damage in the storage of whole grains and polished rice but brown rice (unpolished rice retaining film of oil rich coating) becomes rancid on exposure to air in humid condition.

Minor millets

Several variety of minor millets like ragi or madua or finger miller (Eleusine caracona), sawan or samak (Echinocloa), tagun or kangani or fox tail millet (Setaria italica), kodon or kodra or ditch millet (Paspalum scorbiculatum), proso millet (Panicum miliaceum) and bulbrush millet (Pennisetum americana) etc. These are local feeds of different region and produced in small quantity. These are generally fed to small flocks in backyard and also to pigeons and ornamental birds.

Wheat (Triticum aestivum)

Like rice wheat is another cereal crop of global importance constituting staple food of humans in many countries. A large number of wheat varieties of variable texture and chemical composition are produced in different countries. Feeding of wheat to livestock and poultry was banned for many decades in India and use of second and third grades of wheat has been allowed for the feeding of animals and poultry. Crude protein content ranges from as low as 7% in the native varieties to as high as 22% in few highly improved varieties. In common wheat varieties crude protein content ranges from 10 to 14% on dry matter basis. The production of soft varieties of light brown colour is more common than the hard varieties of dark brown to reddish brown colour. Protein content is affected by the variety and agro-climatic conditions during crop production. The quality of grain for food value is determined by the quality and quantity of protein. The protein of endosperm is gluten which is a mixture of prolamin also known as gliadin and the glutelin or glutenin. Lysine content in glutenin is much higher than the content in gliadin. Gluten content determines the suitability of wheat for the manufacture of ready to eat foods like biscuits and bread because gluten content is responsible for elasticity of the flour.

Wheat flour is not used in the feeds of poultry because it makes the feed pasty which creats problem in swallowing and passage through the alimentary canal. Ground wheat of high gluten contents should not be fed to poultry because it will become pasty and may cause impaction of crop. Aging of wheat kernels for three four months or reconstitution before feeding is required for avoiding the occurrence of digestive disorders. The use of wheat is increasing in the diets of broilers for the production of meat containing white adipose tissue and also solid fat. Broken wheat is generally cheaper than the whole maize.

Barley (Hardeum vulgare)

Metabolizable energy content of whole grain barley is less than the wheat due to presence of 10-15 per cent husk around the kernels. For the feeding of slow growing layer strains, laying hens and foraging birds 30-40 per cent course crushed barley may be used without any adverse effect on performance. But, for the feeding of broilers dehusking or pearl making is required. ME value of whole barley grain for poultry is about 2900 kcal per kg dry matter whereas that of pearl barely is about 3200 kcal per kg DM. Crude protein content ranges from 6 to 12 per cent on DM basis and protein is deficient in many essential amino acids specially the lysine. Feeding of whole grain barley to poultry should not be recommended because the awns present on barley grain may damage the alimentary canal and digestive disorders may occur. It will also have repulsive effect on intake due to sharpness of the awn.

Oats (Avena sativa)

It is not a feed of choice for poultry birds requiring low fibre cereals of high energy value. However, it is satisfactory for the foraging birds like duck and goose. Several variety of oats have been developed but it is a preferred cereal fodder for the ruminants and equines. However, when available it may be fed in small quantity (5-10%). Higher proportion may be used after dehusking the grain which will be an expensive venture. Due to less grain production oat is normally not used for the feeding of poultry specially the broilers. Like many other cereals it is a poor source of protein of inferior quality due to deficient content of many essential amino acids.

Rye (Secale cereale) and Triticale (Wheat x Rye cross)

These crops are rare in India and do not constitute the diets of poultry. However, basic informations are the academic demand. Rye is not a suitable cereal grain for the feeding of poultry due to presence of two antinutritional factors. The bran contains an appetite depressing factor and a growth depressing constituent in the grain. The third harmful factor is exogenous and produced in the seeds due to mould (Claviceps purpurea) infestation which produces ergot.

Triticale is hybrid of wheat and rye developed for disease resistance. This is rarely available in India and has no significance in the feeding of poultry. Although protein quality is better than the wheat and barely, due to higher content of lysine and sulphur containing amino acids but the presence of trypsin inhibitors and alkyl resorcinols for the infestation of ergot producing mold, the Claviceps purpurea.

Cereal grains screenings and shortings

These are generally mixture of undersize grains and seeds of weeds. These are normally not used by the compound feed manufacturing companies because of uncertain availability and highly variable chemical composition. However, these products are quite cheap in the Indian markets and used by backyard poultry keepers for the supplementation of day long scavenging on scrubs.

Damaged grains of food corporation etc.

Every year large quantity of wheat and rice are procured and stored for supply through the public distribution system. Considerable damage occurs annually and ultimately damaged grains of different grades are sold as cattle and poultry feeds. Although, quality of such damaged grains is declared but care must be taken to ascertain the infestation level of fungi, concentration of mycotoxins, if any and the concentration of uric acid which indicates the level of insect damage. It will be always better to assess the presence of insecticides and pesticides often used for protection against insect (tribolium) and rodents (rats and mice).

Cereals milling by-products for poultry feeding

Some of the common cereal milling by-products used in the compounded feeds of poultry are wheat bran, broken rice, rice bran, rice polish, deoiled rice bran (DRB), maize gluten and maize gluten feed. The last two by products contain high percentage of crude protein.

Wheat bran

Wheat bran constitutes 14-17 percent of grain and separated by sieving into fine white flour and bran. It contains 13-16 per cent crude protein and almost equal amount of crude fibre. It is not a good feed for poultry due to bulky nature and low digestibility. ME value of wheat bran for poultry is 1000-1200 kcal per kg DM. Although wheat bran is a richer source of many minerals but their availability is quite low due to phytate formation.

Rice bran

It is energy rich milling by-product of rice obtained from milling and polishing. Rice bran is a mixture of rice polish, embryo and small amount of broken rice. The bran produced by domestic milling also contains fine particles of hulls. High oil content though increases its energy value but significantly reduces keeping quality. Rice bran becomes rancid on exposure to air and process is faster during humid hot climate. Rice bran contains 10-14 per cent crude protein and 11-17 percent oil rich in unsaturated fatty acids. Average ME value of rice polish, rice ban and deoiled rice bran is 2900, 2700 and 2000 kcal per kg dry matter respectively. Deoiling increases dustiness but it can be corrected by mixing molasses. Rice hull is unfit for poultry feeding.

Miscellenious energy feeds

Some nontraditional feeds for energy supply in the diets of poultry are molasses, cassava, potato, sweet potato, beet root, turnip etc.

Molasses

It is sugar rich by product of sugarcane or beet root sugar production industries. Small amount of sugarcane molasses is used for reducing the dustiness of compounded feeds containing deoiled rice bran. Soluble sugar content is more than 40 percent, moisture content is less than 20 per cent, protein is about 3 percent and total ash is 15-25 per cent. ME value of sugar cane molasses ranges from 2300 to 2450 kacal per kg DM. It is mixed upto 10% in poultry feeds. Beet molasses is not suitable for the feeding of poultry.

Potato (Solanum tuberosum)

Potato shortings include small size, cut tubers and residue left during the manufacture of products like chips and finger chips etc. The raw potato contains harmful antinutritional and toxic substances which are destroyed by heat treatment. Potatoes contain more than 79 percent starch, less than 4 percent crude fibre and 0.9 per cent ether extract. Crude protein ranges from 7 to 12 percent in different varieties. Raw tubers particularly sun exposed green tubers and sprouted tubers contain higher concentration of toxic compounds. Almost half of the nitrogenous constituents of crude protein are non protein compounds, alkaloids and glycoalkaloids. The alkaloid present in free form is solanidine and the glycol-alkaloids are solanine and chaconine. ME value of potatoes for poultry is comparable with maize.

For the feeding of poultry and other farm animals the potatoes should be boiled or baked. The water should be drained from the boiled tubers. After this these are dried and ground for incorporation in the poultry feeds. In case of densified feed preparation entire maize may be replaced without any significant adverse effect on the health and production.

Sweet potatoes (Ipomoea batata)

The two common varieties of sweet potatoes are the white tubers and red peel tubers containing pale yellow pulp. The later is rich in carotenoids. Sweet potatoes are similar to potatoes in energy value but contain less protein. The harmful compound present in sweet potatoes is trypsin inhibitor which is not destroyed by sun drying and requires cooking or rosting before feeding or drying for storage.

Trapa (Trapa bispinosa)

It is an aquatic crop of hot-humid tropic. The fruits are rich sources of energy comparable with potatotes and cereal grains. Mature crop contains more than 18 percent dry matter. The fruits are dried, peeled and coarse ground for feeding. The importance of dried trapa fruits has increased in India due to increased consumption in India and neighbouring countries as it is a non cereal food and consumed by people observing fast on religious occasions.

Cassava or Tapioca (Manihot esculenta)

It is a tropical tuber crop produced in the hot-humid areas. The tubers are elongated with rough surface. It contains a thin fibrous central core. Cassava is used for the preparation of flour, starch and globules like that of sago. Heat treatment of tuber is necessary for the destruction and inactivation of two toxic constituents, the cyanogenic glucosides namely linamarin and lotstralin. These cyanogenic glucosides liberate toxic hydrocyanic acid on the action of digestive juices. For the removal of cyanogenic glycosides the tubers are properly boiled and water is squeezed before eating. The boiled cassava tubers are dried and ground for feeding. ME value of cassava is comparable with potatoes and can be used to replace part of cereal grains from the compounded feeds of poultry. Care is taken to balance the diets with protein supplements.

Non-Conventional energy feeds

Some non-conventional energy feeds like sal seed meal, mahua seed cake, castor bean meal, nahor seed meal, azar seed meal, tamarind seed meal, mango seed kernel cake and neem seek kernel cake have been used after processing in very small quantity in the compounded feeds of poultry. However, so far no break through has been confirmed. The claim of utilization of sal seed meal at 2.5 to 7.5 per cent level in the diets of growing and laying birds of layer strains was found as filler on recalculation of results by the author and his colleague during early years of 1980s. Almost all such feeds contain one or more incriminating factors. The processing methods so far developed are expensive, labour intensive and uneconomical. Thus, at the present status of knowledge it will not be wise step to suggest the use of non-conventional feeds for commercial poultry production.

For the economical use of agro-industrial by products wholestic approach is required. No doubt that there is shortage of food grains on calculation basis. At the same time huge quantity of wheat and rice procured by the governments is damaged every year. Liberty should be given to livestock and poultry feed manufacturing companies to purchase cereal grains directly from the producers. In this process farmers will also get better price and food grains will also be not damaged by climatic and biological exposers. There is also shortage of organic manure due to which fertility of arable land is decreasing. Indian soil is facing a serious problem of organic carbon deficit which is an essential constituents of plants and from plants in the form of carbohydrates, proteins and fats it is transferred to animal and human body for body tissue development.

In a very long period of more than half a century research huge amount of money has been used for the development of suitable and economically viable technologies for the utilization of these agro-industrial by products. The out put may not be considered commensurate with the expenditure and precious time of scientists and technical personnels alongwith chemicals and laboratory facilities. It may be a more practical and useful approach to use these organic by products for indirect contribution in food grain production by increasing the yield (Fig. 2.1).

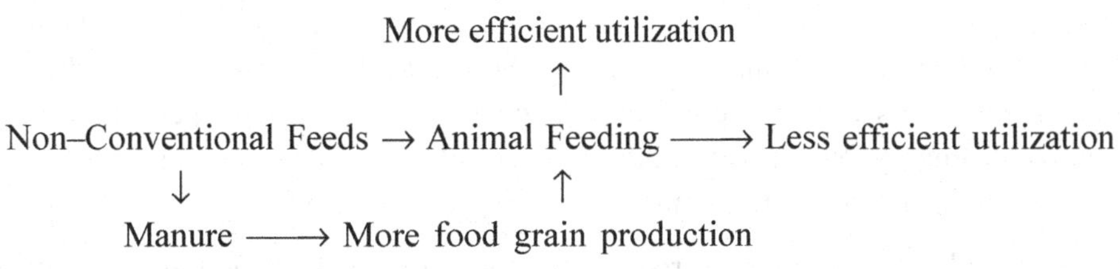

Fig. 2.1: Scheme for utilization of agro-industrial by products

Fat and oils

Fat and oils are the condensed sources of energy capable to supply about 2.25 times more energy than the carbohydrates and proteins. However, there is limitation of utilization and very high level of fats and oils in diets will not be utilized efficiently by the avian species excluding the carnivorous and piscivorous birds. Total lipids in the standard balanced feeds of poultry should be about 5 per cent. A lower level below 3 percent may have adverse effects on production and reproduction. Very high content beyond 10 percent will reduce the digestibility of other nutrients and may also have adverse effects on performance of the poultry birds. However, this is not applicable to carnivorous species like vultures, hawk and Owls etc. Deficient supply of fat or oil will also produce deficiency of linoleic acid and the fat soluble vitamins.

2. PROTEIN RICH FEED INGREDIENTS

Dietary requirement of protein is high for the poultry. The quantity and quality both are important because poultry birds are not able to synthesize about 10 amino acids to meet their optimum physiological requirements for growth, production and maintenance. These are called essential amino acids or indispensable amino acids. Perhaps there is no single proteinous feed to meet the optimum requirements of essential amino acids for supporting different performances. Therefore, amino acid composition of proteinous feeds is also required for selection of combinations of two or more sources of protein for providing balanced amino acids. Proteinous feeds of animal origin are rich sources of essential amino acids and small amount supplementation is found to balance the amino acid composition in the compounded feeds of poultry. Nutritional characteristics of some common and also non-conventional sources of protein feeds have been described briefly for help in the selection of protein concentrates for preparation of feeds of different classes of poultry birds.

A. Oil seed cakes

Earlier three types of oilseed cakes were produced in the Indian sub-continent. There were kolhu pressed cakes containing about 20-30 percent unextracted oil residue. These cakes were more palatable and contained higher concentration of metabolizable energy. Now a days kolhu pressed oil cake is a rare commodity and has no role in commercial feed manufacture.

The second category is comprised of expeller or machine pressed oil seed cakes. In addition to protein content the expeller pressed oil seed cakes are good sources of available energy due to moderate presence of 5-6 percent unextracted oil in the cakes. Protein content is more than the kolhu pressed cake and less than the solvent extracted cake of the same oil seed. The cake is highly palatable and available in the market of many areas.

The third category is that of solvent extracted oil seed cake or meal containing less than 1 per cent unextracted residual oil due to which many oilseed cakes become dusty on grinding. Solvent extracted oil seed cakes now dominate the market and extensively used for the manufacturing of compounded complete feeds of poultry and supplements of livestock.

Some of the oil seeds are contained in a hard coat known as testa or cortex. This considerably reduces the nutritive value of oilseed cakes. Therefore, for the production of good quality cakes such seeds are partially or completely decorticated. Some of the common oilseeds with hard cortex are cotton seed and sun flower seed. The groundnut is contained in a fibrous shell of low palatability and digestibility. Removal of hulls and shell known as decortication decreases the fibre content and significantly improves the palatability, digestibility and nutritive value of decorticated oil seed cakes.

The nitrogenous constituents of the oil seeds and oil seed cake are greatly true protein comprising more than 95 per cent of the total nitrogenous components. Total protein content and biological values of animal protein sources like egg protein, milk protein, fish protein and meat protein are higher than most of the plant proteins. The biological values of protein feeds determined for rats and mice are generally applicable to mammalian species but rarely to avian species because arginine and glycine are required in the feeds for tissue synthesis of poultry. The digestibility of standard common feeds of poultry is fairly comparable and crude protein value alongwith amino acids composition may be a good tool for the selection of suitable cereal and protein supplement combination for balancing the amino acids in compounded feeds. Three main essential amino acids found limiting in poultry production are lysine, methionine and tryptophane. It has been found that compounded diets of poultry meeting the requirements of these three amino acids will be containing adequate quantity of remaining essential amino acids. Therefore, at least these three main limiting amino acids should be given due consideration in the selection of feeds, protein supplements and compounding of diets for poultry.

Plant protein Feeds

Soyabean oilseed meal and groundnut (peanut) seed meal are the common protenous feeds of plant sources which are mostly used in the preparation of poultry feeds. The other vegetable proteion sources are decorticated cotton seed cake, rape seed mustard seed cake, sesame (til or gingili) cake and linseed or flax cake. Some less common protein sources are niger (ram-til) cake, coconut (Copra) meal, sunflower seed cake, safflower seed cake, guar meal, maize gluten, maize gluten feed and palm kernel meal. A few non-conventional proteinous feed not yet fully accepted for the feeding of poultry are need seed kernel cake, castor bean meal and shortings of pulses. Small amount of brewers' yeast and torula yeast is also used for balancing the amino acids content in the feeds.

Plant protein concentrates specially kolhu (ghani) pressed and expeller pressed oilseed cakes containing 5 to 10 per cent oil also contribute significantly to the energy content of compounded poultry feeds.

Soyabean (Glycine max) meal

With the increase in production the use of soyabean meal and whole soyabean or full fat soyabean meal has increased in the diets of poultry, pigs and dairy animals in India. However, raw soyabean seed contains many toxic and antinutrition constituents that are also present in the soyabean meals prepared without proper processing. These harmful constituents are allergens, anticoagulants, goitrogen and trypsin (protease) inhibitors. Out of six protease inhibitors in raw soyabean two have been found more detrimental. These are the Bowman-Birk chymotrypsin inhibitor and the Kunitz antitrypsin factor. Another anti-nutritional factor in raw soyabean is a phytoestrogen, genistein in high concentration. Almost all harmful substances in the raw soyabean are heat labile. Therefore, proper heat treatment, toasting or cooking is necessary before the incorporation of soyabean meal in diets of poultry and pigs.

Soyabean meal is a rich source of highly digestible protein. Crude protein content ranges from 40 to 50 per cent of dry matter. Higher values and higher digestibility are observed in decorticated soyabean meal. Moreover, this feed is deficient in water soluble vitamins of B-complex group and maize-soyabean diet supplemented with methionine has been found very good diet for the fast growing broilers.

Soyabean meal is quite popular in USA and its production and consumption is also increasing in India. The national soyabean processors association (NSPA) in conjunction with the Association of American feed control officials (AAFCO) and the international feed number (IFN) have specified standards for different types of soybean meal produced and marketed. The definitions adopted by these organizations for different kinds of soyabean extraction products may be summarized as follows.

A. **Expeller soybean meal, soybean cake:** The solid residue left after the extraction of oil from soybean seeds. Oil may be extracted by pressure or solvent extraction. The cake is graded according to crude protein content.

B. Soybean flakes include soybean meal containing minimum 44% protein on DM basis and dehulled soybean cake containing minimum 48% protein on DM basis. These two products are produced by cracking, heating and flaking soybeans. The oil is extracted by a organic solvent like hexane. The residue after oil extraction (flakes) is cooked and ground to make meal. There should not be detectable seeds of any weed. A nontoxic conditioning substance may be used for preventing cake formation and declaration of name of conditioning agent is mandatory.

For the international trade it is a common practice to use the term PROFAT SOYBEAN MEAL (PFSM) which shows the total content of protein and fat in the product. Fat content in profat soybean meal will be less than 4.5% of dry matter and may range from 0.5 to 4.5%. The protein and oil combination in a 44.5% profat soybean meal may be any of the following composition.

1. Protein 44% and oil 0.5% is 44.5% profat soybean meal.
2. Protein 43% and oil 1.5% is a 43.0% profit soybean meal.
3. Protein 42% and oil 2.5% is a 42.0% profat soybean
4. Protein 41.5% and oil 3% is a 41.5 % profat soybean meal.
5. Protein 40% and oil 4.5 % is a 40.0 % profat soybean meal.

Therefore, care should be taken to spell the protein content during purchase of soybean meal because in many countries price of protein concentrates including soybean meal is fixed on the basis of intact protein content in the feed.

Groundnut or peanut (Arachis hypogea) cake

Groundnut cake particularly decorticated groundnut cake is also used extensively for the manufacture of poultry feeds. The types of groundnut cakes produced in India are now largely decorticated solvent extracted cake of less than 0.9% fat content and higher than 45% protein content. The digestibility is high but content of some important essential amino acids is quite low. These are lysine and methionine. Groundnut kernel and cake are highly susceptible to Aspergillus infestation and aflatoxins are serious limiting factors. Since production and availability of soyabean meal is increasing the use of groundnut cake in the diets of poultry is decreasing. A cake prepared from the healthy kernels by solvent extraction method is more palatable and nutritious. Groundnut oil is highly susceptible to oxidation causing rancidity due to which keeping quality is low and self life is short. Incidence of adultration of groundnut cake has increased in recent years which has adverse effect on the quality and feeding value of the groundnut cake.

Sesame or til or gingili cake (Sesamum indicum)

Sesame cake is produced in some parts of India and used by small feed manufacturers for the compounding of poultry feeds. Both expeller pressed and solvent extracted cake containing about 40% and 50% crude protein respectively are available in the market. Sesame cake protein is rich in arginine, leucine and methioinine but low in lysine. The residual oil in expeller pressed cake is unsaturated due to which it is not preferred in the diets of broilers. Hulls of sesame seeds contain high concentration of oxalates due to which only decorticated sesame seed cake is used in the diets of poultry. Considerable quantity of whole sesame seed and decorticated sesame seed is also used for the preparation of food items for human consumption. Sesame meal is laxative and it is normally not used in the diets of starter and grower chicks but may be used upto 15 per cent in layers feed. Presence of phytic acid significantly affect the availability of minerals. Therefore, heat treatment is necessary for the inactivation of phytic acid for increasing the availability of minerals from the cake.

Rape seed-mustard seed (Brassica spp.) cake

Several varieties of black, yellow and brown rape seed- mustard seed are extracted by pressure and solvent extraction. The expeller pressed cake contains 4-6% oil and organic solvent extracted cake contains less than 0.9% oil. The antinutritional harmful substances in common rape seed mustard seed cake are glucosinolates and mirosinase (thioglucosidase). These substances are broken down to harmful products like isothiocynates, organic thiocyanates, nitrites and goitrin known as 5-vinyloxazolidine 2-thione. These are responsible for goiter formation and damage of liver and kidney tissues. Another harmful factor is erusic acid responsible for damaging heart but its residue in cake is minimal and has insignificant harmful effect. It also contains variable amount of tannins. Only lysine content is less among the essential amino acids in the cake. Incorporation of even 10% cake in the diets of layers specially those laying brown shelled eggs it may impart fishy taints because these strains of fowl are unable to oxidize trimethylamine produced from sinapine which is a polyphenolic choline ester. Therefore, rape seed mustard seed cake should be used at low level only when other protein feeds suitable for poultry are not available.

In Canada, zero and double zero varieties of rape seed-mustard seed (Canola seed) containing highly reduced amount of erusic acid and glucosinolates have been developed. Production of this variety will increase in many country and it may be used in the compounded feeds of poultry to considerable levels.

Cotton seed (Gossypium spp.) cake

Common cotton seed cake contains high fibre (15-20%) and low protein (20-25%) but decorticated cake contains less than 9% crude fibre and 40-46% crude protein in dry matter. Gossypol is a toxic constituent which is present in free and bound forms. Free gossypol is more harmful. Gossypol is an antioxidant and reduces oxygen uptake by the red blood cells and transportation. Cotton seed cake content should not be more than 10% in compounded poultry feeds and free gossypol should not exceed 100 mg per kg diet. Cotton seed cake protein is deficient in lysine,methionine, tryptophan and threomine. This cake needs careful storage for protection against Aspergilus infestation the source of aflatoxins. Cotton seed cake is not a common feed for poultry and used during shortage of common protein supplements.

Sunflower seed (Alianthus) cake

Crude fibre content is very high on whole seed extraction which is more than 25% (20-30%) and crude protein content is 25-30%. Decortication reduces fibre and increases protein content. Although essential amino acids content is better than many cakes but ME value is low. Digestibility of cake is also low due to the presence of chlorogenic acids. It is rarely used in the compounded feeds of poultry.

Safflower (Carthamus tinctorius) seed cake

Safflower or kardi or barrey seed meal is produced in parts of Madhya Pradesh, Gujarat, Rajsthan, Karnatak, Andhra Pradesh and small areas of Uttar Pradesh and Bihar. The ridged seeds are elongated, pentagonal and white hard coated. It contains a bitter principle probably a saponin. Whole seed cake contains less than 30% protein while decorticated solvent extracted meal may contain up to 55% protein. The oil is highly unsaturated. The limiting amino acids are lysine and methionine. It is not a preferred protein supplement for poultry but may be used upto 5-6% in layers diet. For increasing the level of unsaturated (Omega) fatty acids in eggs. Expeller pressed cake (7-10% oil) may be used in layers feeds.

Nigar or Ram til seed (Guizotia abyssinica) meal

The niger cake contains about 35% crude protein, 2% oil, 13% crude fibre and 2290 Kcal ME per kg dry matter. The feed is less palatable. There appears to be some antinutritional factors in the niger seed cake and even 10 per cent level in compounded feeds of starter chicks has growth depressing effect. However, small amount can be incorporated alongwith palatable feeds and fish meal.

Coconut (Cocos nucifera) cake

Mostly expeller pressed meal containing about 15% protein, 8-10% oil and 11-12 % fibre or solvent extracted meal of 20-25% protein, about 1% oil and 14-16% fibre are available. High fat cake has short self life due to development of rancidity caused by oxidation of oil. The ME value is low and protein quality is poor due to deficient content of lysine and histidine for poultry and other simple stomached animals. The cake is highly susceptible for mould growth resulting in aflatoxins production. The feed is not a suitable protein concentrate for feeding the poultry. However, due to easy availability small amount is used in the diets of layer strains.

Palm kernel (Eleis quinensis) cake

It is poor quality protein supplement for poultry feeding. Protein content is low, digestibility is less and lysine is deficient for poultry. Solvent extracted meal is not palatable and fibrous. The feed is not liked by poultry. However, in scarcity condition it can be fed after mixing with other palatable feeds in bind form like pellet or flake etc.

Linseed or Flax seed (Linum usitatissimum) cake

It is not suitable for any class of poultry bird and as far as possible linseed or linseed meal should not be incorporated in the compounded feeds of poultry. Mucilage in linseed cake is indigestible. The toxic substances are linamarin and linase. The later hydrolyses linamarin to produce highly toxic hydrogen cyanide. As little as 5% linseed meal in the diets of chicks suppresses growth. Protein quality is low due to deficient content of lysine and methionine. Death in turkey poults occurs on feeding 10% untreated linseed meal in diet. An antipyridoxine factor is considered responsible for increasing the requirement of pyridoxine (Vitamin B6) in the diet. Heat treatment improves the nutritional quality by destroying the enzyme linase and also the anti-pyridoxine factor. Feeding of linseed meal to poultry should be avoided.

Protein rich maize kernel milling by-products

In the process of starch and oil extraction from the maize (Zea mays) kernel and maize germ the by products produced are maize bran, maize gluten and maize germ or embryo. The three are either separated into bran, gluten and germ or used mixed for animal feeding. Maize germ is a very rich source of unsaturated omega fatty acids and extracted for human consumption. The common maize by product feeds available in considerable quantity for animal feeding are maize gluten meal, maize gluten feed and maize cake or maize germ meal. Crude protein content in maize gluten ranges from 50 to 60 percent which is more than only 20 to 25 % in maize gluten feed. The later contains higher (6-8%) level of crude fibre. Maize gluten is a rich source of methionine. Inclusion of 3 to 5 percent maize gluten in maize soyabean feed meets the requirement of methionine. Similarly in double (6-10%) ratio maize gluten feed meets the requirement of methionine. These feeds are deficient in lysine and arginine. These feeds obtained from the milling of yellow and orange maize are rich in xanthophylls. Maize germ contains high concentration of residual oil in expeller pressed product which has short keeping quality and spoiled by development of rancidity.

Guar meal (Cymopsis tetragonaloba)

The by product left after the extraction of gum from guar (Cluster bean) seed is known as guar meal. It is a good source of lysine and methionine. Protein content is more than 40 percent but quality is hampered by the presence of harmful antinutritional components like trypsin inhibitors, urease and haemagglutins. These are destroyed or inactivated by moist heat treatment or autoclaving. It can be used at 3-5 percent level in otherwise balanced feeds of poultry.

Leguminous seeds as protein source for poultry

Due to shortage of conventional protein feeds for poultry a successful attempt was made to supply dietary protein requirement of chicken through a mixture of leguminous seeds (K.K. Baruah, A. Saikia and N.N. Pathak, 1975). The results indicated scope for the development of such feeds using small amount of good quality fish meal.

Leguminosae is a big family of more than 12000 plants of three subfamilies and four tribes. The sub families are papilionacae, cicilpinacae and mimosae. This family is also differentiated into four tribes, i.e. the hedysareae, the vicinae, the phaseolae and the genisteae. The common genera of food value of different tribes are presented in Table 2.1.

Table 2.1: Common genera of food value in four tribes of leguminosae family

Tribe	Common genera
Hedysareae	Groundnut (Arachis hypogia)
Vicineae	Vicia spp., Cicer spp., Lathyrus spp, Lens spp. Pisum spp.
Phaseoleae	Phaseolus spp., Dolichos spp, Glycine spp.
Genisteae	Lupinus spp.

Most of the leguminous seeds contain toxic and antinutritional factors (Table 2.2) and require proper treatments to make feeds fit for feeding. The common treatments are cooking, boiling, autoclaving and roasting. Lathyrism caused by a toxic component lethyrisine (beta-aminopropionatrile) in seed. (Table 2.2).

Table 2.2: Toxic and harmful substances in leguminous foods.

Name of legume	Toxic and antinutritional factors
Lathyrus sativus	Beta-aminopropionitrile
Glycine max	Trypsin inhibitor
Phaseolus lunate	Phaseolunatin (a cyanogenic glucoside), lectins.
Phaseolus vulgaris	Lectins (these are generally heat labile)
Cicer ensiformis	
Dolichus biflorus	
Dolichus lablab	

Beans and peas are moderate source of protein (18-24%) and satisfactory energy for poultry. Cystine and methionine are the limiting amino acids for poultry. Pea, cowpea, gram and other pulses can be used upto 10 per cent in layers feed after proper treatment for the inactivation of antinutritional factors. The cost of leguminous foods has highly increased during last two decades and there appears to be no scope of using seeds of beans and peas in the diets of poultry and animals.

Some non-conventional protein feeds

Castor bean (*Ricinus communis*) and neem seed (Azadirchta indica) kernel cake have been claimed suitable for the feeding of poultry birds. However, the cost of processing, method of processing and problem of disposal of affluents are the limiting factors for commercial use in poultry feed. It can be recommended for feeding to poultry only when other feeds are not available.

Ambadi (Hibiscus cannabinus) seed meal

It is a medium protein (26-30%) and high fibre (20-24%) feed. The palatability and digestibility are low. Protein is deficient in lysine and methionine. It is normally not incorporated in the compounded feeds of poultry.

Karanj (Pangamia glabra) seed meal

Solvent extracted decorticated karanj seed cake contains less than 1% oil and more than 30 percent protein. But contains toxic factors like karanzin. Expeller pressed karanj cake is not palatable and harmful but solvent extracted karanj seed cake containing less than1% oil and 7% crude fibre can be used upto 5% in the compounded feeds of poultry. Karanj seed cake is a good source of many essential amino acids.

Rubber (Hevea brasiliensis) seed cake

Rubber plantation is more in Kerala, Karnataka and Tamil Nadu. Annual production of rubber seed is about 8 to 10 quintals per hectare. Rubber seed kernel cake contains about 30% crude protein. The harmful component in rubber seed is a cyanogenetic glucoside but concentration is generally very low. Rubber seed kernel as well as cake can be fed to poultry reared by small holders in backyard management. Decorticated solvent extracted rubber seed kernel cake can be incorporated upto 20 per cent in the compounded feeds of layers. It is a satisfactory source of essential amino acids lysine, methionine and threonine.

Neem (Azadirachta indica) seed kernel cake

Now solvent extracted neem seed kernel cake of high protein and little oil and fibre content is produced. It contains bitter principles nimbidin and nimbin, which make the cake unpalatable. Treated neem seed kernel cake is palatable and can be incorporated in the diets of poultry upto 20 percent without considerable effect on productivity. The processing cost is quite high and neem seed cake has other important uses in agriculture and horticulture, which are also important for food supply. Therefore, it may be a better approach to use neem seed cake as organic manure and its water extract for the control of insects on the crops and animals.

Dehydrated leaf meal

Protein and carotenoid rich leaves of leguminous crops like berseem, lucerne, cow pea and cluster bean are dehydrated in shade or drier to reduce the moisture content below 12 per cent. The dehydrated leaves are coarse ground to make leaf meal. These are palatable and improves the appearance of compounded feeds. Crude protein content varies from 25-32 percent of dry matter. Leaf meals are also good sources of green and yellow colour of carotenoids. Digestibility in poultry is about 70 percent and energy value is less than the cereal grains. These can be incorporated upto 5 percent in growers diets and 10 percent in layers feeds. Production of leaf meal in India is probably not practiced.

Animal Protein feeds

Since poultry birds are unable to synthesize about 11 amino acids essential for normal growth and production, It is necessary to provide feeds for balancing the nutritional requirements for various physiological functions. These are 10 determined for the rat, i.e. AVHILLMPTT or arginine, valine, histidine, isoleucine, leucine, lysine, methionine, phenylalanine, threonine and tryptophane. The additional one is glyucine. Almost all vegetable proteion feeds are deficient in few essential amino acids specially the lysine, methionine and tryptophane. Animal protein feeds are rich source of essential amino acids and required incorporation in small amount for balancing the requirements of essential amino acids of poultry. Among the available sources of animal protein feeds, fish meal is the most popular ingredient in India.

Fish meal

Fish meal is the most common protein supplement used for balancing the essential amino acids content and ratios in the compounded poultry feeds. Three types of fish meals are available in the Indian markets. These are whole dried fish or jawla fish, ground fish meal and fish trash meal containing 50-70, 40-55 and 20-40 percent protein on dry matter basis respectively. The last category contains more than 25 percent total ash containing high proportion of sand, and use of such fish meal may have adverse effects on the production of poultry. Fish meal is a very good source of all essential amino acids except the tryptophane. The other important nutrients in good quality fish meal are minerals like calcium, phosphorus and iodine and vitamins like vitamin B_{12}, choline, riboflavin and pantothenic acid. Fishy odour is transferable in the products and fish meal should not constitute higher than 10 percent of the complete feed. Fish meal generally contains high concentration (1 to 6% or more) of common salt. The salt content in fish meal should be taken into account while formulating the diets so that level of common salt does not exceed 0.5% of feeds on dry matter basis.

Meat meal

It is another important source of animal protein for the feeding of poultry. Meat meal is prepared from the discarded carcass unfit for human consumption and unspoiled (unautolysed) carcass of fallen animals. The flesh is cooked/pressure cooked/autoclaved, dried and ground. Fat is generally extracted. Meat meal contains 50 to 80% protein and high level of lysine but sulphur amino acids are low.

Meat cum bone meal

This protenious feed is prepared from the whole carcass after flaying skin and removing horns, hooves and digesta from the alimentary canal. The quality of meat –cum-bone meal is highly variable and depends on the body condition of the animal. This product is used only when fish meat or meat meal is not available.

Blood meal

It is a low quality protein due to poor palatability and low digestibility. Blood meal is prepared from the clot available after the separation of serum or plasma. The clot is mixed with equal amount of deoiled rice bran or wheat bran, cooked and ground.

Silk worm pupae meal

It is prepared from the deoiled silkworm pupae, which are dried and ground. Deoiled silkworm pupae meal is a rich source of good quality protein. Amino acid composition is very good and feed contains good amount of lysine, methionine, tryptophane, isoleucine and arginine but threonine content is deficient. Fresh silkworm pupae are also used in human dietary of some tribal community after cooking.

Hatchery by-product meal

Hatchery by product meal is prepared from unhatched eggs, yolk bound dead embryo and empty shells. It is a good source of protein and energy and rich source of calcium. At large hatchery some times shells are separated and processed for mineral mixture preparation. Shell free hatchery by product meal is a rich source of good quality protein and metabolizable energy.

Chicken processing residue meal

Various shortings and by products available at chicken dressing plants are legs below knee, giblet free internal organs, head, blood and feathers. The feathers are separated and remaining by products are cooked, dried and ground. It is a good source of protein and energy.

Feather meal

After shorting soft feather for cushion making and long feathers (sickles) for the preparation of decoration items the remaining feather is macerated, hydrolyzed, dried and ground to make feather meal. Although feather meal contains about (60-93) 85% protein but essential amino acids lysine, methonine and trypophane are less.

The feed is less palatable and less digestible. Pressure cooking for longer duration may increase digestibility but palatability is little affected. Feather meal is rarely used and that too at very low level (2-5%) in the diets of poultry.

Skim milk powder waste

During preparation and handling some amount of skim milk powder is soiled and used for the feeding of animals. Protein content is about 35% in cow milk and 40% in buffalo milk powder. Skim milk powder is a good source of essential amino acids and several minerals but poor source of iron and fat soluble vitamins. Incorporation in mash has less feed value and for better utilization pelleting is required. It is deficient in cysteine content. The digestibility and biological value of spray dried product is higher than the roller dried product due to partial denaturation at higher temperature of roller drying process.

Single cell protein

Various strains of yeast candida and saccharomyces and bacteria Pseudomonas methylica are cultured on hydrocarbons (gas oil, paraffin), alcohol and molasses. Crude protein in different types of single cell protein ranges from 38 to 82% on dry matter basis. Nucleic acid concentration ranges from 5-12% in yeast and 8-15% in bacterial mass on dry matter basis. About 2 to 5 per cent single cell protein may be used in the poultry diet.

Miscelleneous proteinous feeds

Small amount of different sources of proteins are used by the small scale local feed manufacturers. These are frog trash meal, snail meal and liver residue meal.

Frog trash meal

In the frog leg production industry other flesh and bones are utilized for the production of animal feeds. The residue after removal of hind legs at lumbar region is cooked, dried and ground to make frog trash meal. It is not available for use in large manufacturing companies.

Snail meal

Small variety of marine and terrestrial snails are eaten by foraging semi-aquatic birds like duck, goose and swan. Snails are the sources of good quality protein and minerals.

Liver residue meal

This is a protein and vitamin rich feed available after the extraction of vitamins and oils marketed for medicinal uses as liver extract.

A tentative guide for the use of feeds in compounded feeds of poultry

The nutritional requirements of the very fast growing broiler chicken is much higher than the requirements of laying strains of fowl, guinea fowl, duck and geese. The large size foraging birds like goose, swan, emu and ostrich are capable of extracting their nutritional requirements from fibrous feeds like whole grain paddy, barley, oats and other millets. Therefore, there is need of having a reckner for the selection of feeds for the manufacture of complete balanced diets for growing- finishing broilers and starter, grower and layer chicken. The former is also applicable for the fattening quail and later are applicable for the guinea fowl and intensive reared ducks. Figures provided in table 2.3 may be used for the selection of energy feed and table 2.4 for the selection of protein feeds.

Some products used for improving feeds

In the absence of yellow or red maize in the diet the colour of yolk becomes dull white to pale yellow. Similarly importance of omega fatty acids for cardiac petients and also as protective food the demand of eggs and chicken with high omega 3 fatty acid is increasing. Therefore, feeding of ingredients for the improvement of yolk colour and carotene content in eggs, and omega 3 fatty acids in eggs and chicken following feeds are used.

1. **Fresh green forages:** Fresh leafy fodder like berseem, lucerne, cowpea, sengi, spinach, azolla and lemna are offered to birds after 1-2 hours of offering normal feeds
2. **Leaf meal:** Leaf meal is prepared from leguminous crops like lucerne, berseem, cowpea and forbs like safflower. The leaves are dried in shade or dehydrated in drier and then coarse ground to make meal.
3. Fish oil, linseed oil, safflower oil, maize oil and other sources of oil rich in unsaturated and omega fatty acids are incorporated (about 3%) in the diets of poultry specially the layers.

Mineral supplements

Physiological requirements of most of the dietary essential minerals are not met by the common source of energy and protein feeds and there is need of supplementation of mineral salts for balancing the compounded feeds of different class of poultry. Some of the common sources of different dietary essential minerals are presented in table 2.5 and 2.6.

Table 2.3: Approximate efficient level of energy feeds in compounded feeds of different poultry birds.

Sl.	Name of feed	Starter chick and Broiler	Grower chicks and laying hens	Remarks
1.	Maize	70	70	Yellow variety preferred. Susceptible to fungal infestation (Mycotoxins).
2.	Sorghum/ Jowar	30	60	White and yellow varieties prefereed. Red is rich in tannins.
3.	Wheat	20	30	New grain contains arabino-xylans. Seesoned grain is fed.
4.	Whole grain paddy	–	50 (Duck, goose, swan)	Not suitable for broilers and chicken
5.	Rice white Rice brown Rice broken	10 10 10	25 20 25	Brown rice contains polish and becomes rancid on exposure to moist air.
6.	Finger millet (Ragi or Madua	10	20	After removal of bran (Whole grains may be fed to adults)
7.	Foxtail millet (Cheena)	10	20	After removal of bran (Whole grains may be fed to adults)
8.	Paspalam (Kodon)	5	15	After removal of bran (Whole grains may be fed to adults)
9.	Barley	10	35	High fibre, Beta-glucans present
10.	Pearl millet or bajra	20	60	Tannins present
11.	Rice bran	20	30	Rancidity spoils, Antioxidants are needed.
12.	Deoiled rice bran	–	10	Keeping quality is better.
13.	Wheat bran	5	10	Should be avoided
14.	Oat	–	15	High fibre
15.	Pearled barley	20	40	Husk has been removed
16.	Cooked dry potato	10	30	Compressed feeds.
17.	Tapioca/ cassava	10	20	Should be cooked or autoclaved
18.	Molasses	3	5-10	Loose droppings, soiling of feathers
19.	Bakery by products	5	10-15	Should be free from fungal contamination.
20	Mango seed kernel meal	–	5	Should be mould free

Table 2.4: Level of protein feeds in complete feeds of poultry.

Sl.	Name of feed	Starter chick and Broiler	Grower chicks and laying hens	Remarks
1.	Soyabean meal	30-40	20-30	Methionine supplement needed
2.	Groundnut cake	30-40	20-30	Lysine, methionine and tryptophane deficient
3.	Rape seed-mustard cake	3-4	5-6	Anti-nutritional factors
4.	Decorticated cotton seed cake	10	10	Susceptible for mould growth
5.	Safflower cake	5	10	High fibre
6.	Sunflower seed cake	10	20	High fibre
7.	Sesame cake	10	15	Fibre, phytates and oxalates
8.	Niger cake	5	10	Antinutritional factor
9.	Maize gluten	10	20	Susceptible for mould growth
10.	Maize gluten feed	10	20	Susceptible for mould growth
11	Guar meal	3	5	Heat treatment needed
12.	Ambadi cake	10	20	High fibre
13.	Rubber seed cake	5	10	High fibre, cyanogenetic glucosides
14.	Palm kernel meal	10	20	High fibre
15.	Coconut meal	3	5	High fibre, aflatoxins
	Animal Protein supplements			
16.	Fish meal	10	10	Rancidity, specules, pathogenic microbes
17.	Meat meal	7	5	Pathogenic microbes may be present
18.	Meat cum bone meal	10	7	Pathogenic microbes may be present
19.	Liver residue meal	5	5	Pathogenic contamination may be possible
20.	Silk worm pupae meal	3	3	
21.	Shrimp meal	5	5	
22.	Hatchery waste meal	3	3	Sterilization is necessary
23.	Feather meal	5	3	
24.	Dried yeast	5	3	High nucleic acid content
25.	Skim milk powder	5	3	
26.	Dried distillers residue	–	10	

Table 2.5: Common sources of minerals for poultry

Sl.	Name of product	Mineral content			
1.	Sterilized bone meal	Ca	29%	P	12%
2.	Calcium carbonate (Lime Stone)	Ca	38%	–	–
3.	Dicalcium phosphate	Ca	22%	P	18%
4.	Monocalcium phosphate	Ca	16%	P	21%
5.	Deflorinated rock phosphate	Ca	32%	P	16%
6.	Soft rockphosphate	Ca	17%	P	9%
7.	Oester shell grit	Ca	38%	–	–
8.	Calcium sulphate	Ca	22%	–	
9.	Mono ammonium phosphate	–		P	24%
10.	Diammonium phosphate	–	–	P	20%
11.	Sodium acid phosphate	–		P	22%
12.	Meat-cum bone meal	Ca	10%	P	5%

Table 2.6: Common sources of micro (trace) minerals for poultry

Sl.	Source/Salt		Mineral content %	
1.	Copper sulphate	$CuSO_4.H_2O$	Copper	35
2.	Copper sulphate	$CuSO_4.5H_2O$	Copper	25
3.	Cupric carbonate	$CuCO_3.H_2O$	Copper	53
4.	Cupric oxide	CuO	Copper	75
5.	Ferrous carbonate	$FeCO_3$	Iron	21
6.	Ferrous sulphate	$FeSO_4.\ 7H_2O$	Iron	76
7.	Manganese carbonate	$MnCO_3$	Manganese	47
8.	Manganese sulphate	$MnSO_4.H_2O$	Manganese	25
9.	Manganese sulphate	$MnSO_4.5H_2O$	Manganese	22
10.	Potassium iodate	KIO_3	Iodine	59
11.	Potassium iodide	KI	Iodine	76
12.	Sod. Chloride	$NaCl$	SodiumChlorine	3960
13.	Sod. Selenate	Na_2SeO_4	Seleium	41
14.	Sodium Selenite	Na_2SeO_3	Selenium	45
15.	Zinc Oxide	ZnO	Zinc	73
16.	Zinc sulphate	$ZnSO_4.7H_2O$	Zinc	36

Vitamin supplements

Vitamins are unrelated chemical compounds required in minute quantity for optimum physiological functions.

National research council (NRC, 1994) of USA has suggested 13 vitamins dietary essential. Some vitamins are synthesized in the birds but products are mostly insufficient to meet optimum requirement due to which dietary supplementation is required. Vitamin D is utilized for the production of D3 form or 1, 25-dihydroxycholecalciferol required for the absorption and metabolism of calcium. Vitamin E is a very effective antioxidant in tissues. The other vitamins are mostly a co-factor in enzymic reactions in the body. Choline is not a metabolic catalyst. It is a source of methyl group (CH3) and a constituent of phospholipids responsible for neurotransmission. Niacin is synthesized from tryptophan in liver and vitamin D3 is synthesized from 7-dehydrocholesterol in the skim exposed to light. Some vitamin of B-complex group, say folic acid and B12 are synthesized by normal microorganisms in the caeca and large intestine but these vitamins are available to scavenging birds through coprophagy. Absorption of these vitamins through the walls of caeca and large intestine is not yet clear and needs further exploration. Requirement of vitamin E is increases on the feeding of polyunsaturated fatty acids. Mostly synthetic vitamin sources are used for supplementing the compounded poultry feeds (Table 2.7).

Table 2.7: Sources of vitamin supplements for poultry.

Sl. Vitamin supplement	Vitamin activity IU/g	
Fat soluble vitamins	IU	
1. Vitamin A (Retinol acetate)	500,000 IU/g powder form	
2. Vitamin A and D3 premix	Vitamin A 500,000 IU Vitamin D3 100,000 IU	
3. Vitamin D3	500,000 IU/g powder	
4. Vitamin D3	200,000 IU/g powder	
5. Vitamin E	Vitamin E acetate 50%	Feed grade powder
6. Vitamin E	Vitamin E acetate	Liquid packing
7. Vitamin E	Vitamin K_3	Powder form
Fat and water soluble mixture		
8. Vitamin A B_2 D_3	Commercial products with specification and concentration. Most of these are available in dry powder form for easy storage and handling	
9. Vitamin A B_2 D_3 K		
10. Vitamin A B_2 D_3 E		
Water soluble vitamins		
11. Vitamin B_1		
12. Vitamin B_2		
13. Vitamin B_6		
14. Vitamin B_{12}		
15. Biotin or Vitamin H		
16. Niacin		
17. Choline		

3

NUTRIENTS

The organic and inorganic substances required for the normal function of a living body are known as nutrients. Foods are the main sources of nutrients and plant products form the bulk of diets of herbivores and ommivores. Most of the birds are omnivores and food grains constitute greater part of compounded diets of avian species. The feeds are partitioned into five groups of nutritionally important constituents known as proximate principles. These are crude protein, ether extract, crude fibre, nitrogen free extract (NFE) and ash. Water is an important constituent of most of the feeds and body, and it is essential for the survival and physiological functions of the body and should be counted with nutrients.

For the purpose of poultry feeding there is need of more elaborate composition of feeds which are initially divided into moisture (water) and dry matter. Dry matter of common organic foods includes complex compounds like carbohydrates, proteins, lipids, minerals and vitamins. Birds are simple stomached animals without teeth and cheeks. The morphology of alimentary canal does not provide enough space for the utilization of complex constituents of feeds like fibrous carbohydrates and non protein nitrogenous compounds. Hence, there is need of more detail analysis of food ingredients. The major constituents of common feeds are water, carbohydrates, proteins, lipids, minerals and vitamins.

CARBOHYDRATES

Carbohydrates constitute bulk of the plants and feeds of plant origin in form of sugars and non sugar carbohydrates. In poultry feeding starch is the main and contribution of fibrous carbohydrates (cellulose and hemicellulose) is less and highly variable among the species. Dietary carbohydrates are classified in different manner. However, main consideration in the classification of the carbohydrates is the number of carbon and water in the molecule, size of the molecule and presence of elements or chemical form other than the carbon and water in the molecule as in case of derived carbohydrates.

Cereal grains, leguminous seeds and oil seeds and their milling residues and processing by products are the common sources of carbohydrates for the feeding of poultry, Sugars are rarely used in the diets of poultry but small amount of molasses may be used specially for the pelleting of compounded complete feeds. Classification of carbohydrates is presented in Table. 3.1.

Table 3.1: Classification and sources of common carbohydrates.

Carbohydrates	Monosaccharide constituents	Main sources
A. Monosaccharides		
1. Pentoses ($C_5H_{10}O_5$)		
Arabinose Xylose Ribose	Arabinose Xylose Ribose	Pectic polysaccharides, araban corn cob, wood sugars Nucleic acids
2. Hexoses($C_6H_{12}O_6$)		
Glucose	Glucose	Disaccharides, Polysaccharides
Fructose	Fructose	Sucrose
Galactose	Galactose	Lactose (Milk sugar)
Mannose	Mannore	Polysaccharides
B. Disaccharides ($2C_6H_{12}O_6$)-H_2O or ($C_{12}H_{22}O_{11}$)		
Sucrose	Glucose+fructose	Cane sugar, beet sugar
Maltose	Glucose+glucose	Plants and roots
Lactose	Glucose+galactose	Milk
Cellobiose	Glucose+glucose (glucose – 4-beta glucoside)	Plant fibrous tissue
C. Trisaccharides ($3C_6H_{12}O_6$)-$3H_2O$) or ($C_{18}H_{30}O_{15}$)		
Raffinose	Glucose+fructose +galactose	Cotton Seeds, sugar beets, eucalyptus
Ketose and isoketose	Sucrose+fructose	Plants and seeds of grasses
D. Tetra saccharides ($4C_6H_{12}O_6$)-$4H_2O$ or ($C_{24}H_{40}O_{20}$)		
Stachyose	Polymerized raffinose (galactose + glucose + fructose)	Plants

Disaccharides, trisaccharides and tetrasaccharides together are called oligosaccharides.

[Table Contd.

Contd. Table]

Carbohydrates	Monosaccharide constituents	Main sources
E. Polysaccharides; These are complex carbohydrates of big molucue and non sugar in taste		
(a) Pentosans $(C_5H_{-8}O_4)n$	Pentose	Pentose polymers
1. Araban	Arabinose	Pectins
2. xylan	xylose	Maize cob, wood
(b) Hexosans $(C_6H_{10}O_5)n$	Hexose	Hexose polymers
1. Starch	glucose	Cereals, tubers, roots
2. Dextrin	glucose	Starch
3. Cellulose	glucose	Fibrous part of plants, grains, fruits, roots
4. Glycogen	glucose	Liver and muscles. (Animal starch)
5. Inulin	Fructose	Tubers, potatoes
(c) Mixed polysaccharides		
1. Hemicellulose	Pentoses & hexoses	Fibre parts of platns
2. Pectins	Pentoses & hexoses	Apple, citrus fruits with salts of acids
3. Gums	Pentoses and hexoses	Acacia and other gum oozing trees
F. Derived carbohydrates		
1. Amino sugars Galactosamine	Glucosamine Galactose, CH_3 N	Glucose, CH_3 N Chitin Cartilage
2. Phosphoric acid ester	metabolism cycle	Glucose -1- phosphate phosphoric acid + glucose, Glucose-6- phosphate
3. Deoxy sugars Rhamnose	Deoxy ribose, Ribose Mannose	Nucleic acid Heteropolysaccharides
4. Sugar acids Heteropoly-saccharides Glucose	Aldonic acids (gluconic acid) Aldaric acids	Aldoses Glucose (glucoric acid)
5. Sugar alcohols Sorbitol Galactose Grass silage	 Glucose Mannitol	 Dulctol Mannose and fructose
6. Glycosides Glucoside Fructoside Glycosides	 Glucose Fructose Oligosaccharides polysaccharides	 Cyanogenic compounds

Harmful and toxic carbohydrates

Many harmful and toxic carbohydrates mainly derivatives of glucose, frutose, oligosaccharides and polysaccharides are produced in some plant species. The products of enzymic hydrolysis in the gastrointestinal tract are mainly glucose and hydrocyanic acid (HCN). The additional hydrolytic products of various cynogenic glycosides are shown in Table 3.2.

Table 3.2: Hydrolytic products of cyanogenetic glycosides in addition to glucose and hydrocyanic acid.

Cyanogenetic Glucosides	Additional hydrolytic products	Sources of glycosides
Amygdalin	Benzaldehyde	Kernels of apple, cherries, peach, plum, bitter almond, fruits of rosaceae family
Dhurin	p-hydroxybenzaldehide	Leaves of jowar (Sorghum bicolor, S. vulgare
Linamarin (Phaseolunatin)	Acetone	Linseed or Flax seed, Java beans, cassava
Lotaustralin	Methyl-ethyl ketone	White clover, trefoil
Vicianin	Arabinose, Benzaldehide	Vetch seeds

Functions of carbohydrates in the body

1. Carbohydrates are the major source of dietary metabolizable energy for poultry production.
2. Glycans are universally distributed and found in the cells and inter cellular space. These transmits biochemical signals into the cell and amongst the cells.
3. Contribution of energy from plant carbohydrates is small in carnivorous avian species like owls, hawks and vulturs etc.
4. Carbohydrates are not essential for chicken and they can survive on a dietary energy of lipids and proteins provided protein: energy (ME) ratio is optimum and the sources of fat are triglycerides. In the absence of triglycerides gluconeogenesis from amino acids fails due to non availability of glycerol a product of hydrolysis of triglycerides.
5. Blood glucose level should be maintained for normal body functions. The source may be glucogenesis from food or gluconeogenesis from fat and/ or protein.
6. Participates in cell functions for protein and vitamins synthesis.
7. Indigestible fibrous carbohydrates are hygroscopic and facilitate defaecation.

Role of soluble non-starch polysaccharides (NSPs)

Higher concentration of soluble non starch polysaccharides is not desirable in the diets of poultry due to following reasons:

1. Soluble non starch polysaccharides like pentosans increase viscosity of digesta resulting in higher rate of passage and lowering of digestion of nutrients and duration of absorption of nutrients.
2. These are indigestible in poultry and occupy the space of digestible constituents of the feedstuffs.
3. The performances of birds decrease due to reduced availability of nutrients.
4. These are, therefore, considered anti nutritional factors.

PROTEINS

Proteins are the constitutional tissues responsible for the development of characteristic shape and size of body in conjunction with minerals. Proteins constitute major part of body and their concentration is maximum in muscles, glands and body fluids. These are complex compounds made of carbon, oxygen, hydrogen and nitrogen. Sulphur and some other substances are also required for the synthesis of special proteins. Proteins are synthesized in all living cells and participate in all activities of cells necessary for the maintenance of life.

Protein content in body

Total protein content in the body of animals including birds ranges from 15 to 20 per cent. Body protein content on fat free basis is almost constant which is 21.6 per cent. The ratio of protein in the body decreases with the increase in body fat proportion. In case of poultry larger part of protein is utilized for the synthesis of feathers.

Constituents of proteins

Amino acids are the units required for the synthesis of proteins. Body protein is made of amino acids. Of the 20 amino acids involved in the synthesis of body tissue 10 were found dietary essential because animals are unable to synthesize these amino acids or can synthesize few in very small quantity, much less than the optimum requirements. These amino acids are essential dietary supply and called essential amino acids (EAA) or non-dispensable amino acids.

Classification of proteins

Proteins are broadly classified into three groups on the basis of their physical, chemical and functional properties (Table 3.3).

Table 3.3: Classification of proteins

Globular proteins	Fibrous proteins	Conjugated proteins
Albumin	Collagen	Glycoproteins
Globulins	Elastin	Lipoproteins
Histones	Keratin	Mucoproteins Nucleoproteins Phosphoproteins Protamines

Amino acids

Amino acids required by poultry or any other animal are differentiated into two main groups of non-essential or dispensable amino acids and essential or indispensable amino acids . Dispensable amino acids are synthesized in the cells to meet the physiological requirements of the body on the availability of nitrogenous moiety. Dietary supplementation of essential amino acids is must for the optimum physiological functions. Protein nutrition is very sensitive in poultry production and deficiency or imbalance of few essential amino acids can disturb the production performances of the birds. The essential amino acids of poultry are further distinguished into critical amino acids and limiting amino acids (Table 3.4).

Table 3.4: Amino acids in poultry nutrition

Non essential amino acids	Essential or indispensable amino acids		
	All amino acids	Critical amino acids	Limiting amino acids
Alanine	Lysine	Lysine	Lysine
Aspartic acid	Methionine or	Methionine or	Methionine or
Glutamic acid	Methionine+Cystine	Methionine+Cystine	Methionine+Cystine
Hydrooxyprolin	Tryptophane	Tryptophane	
Proline	Threonine	Threonine	
Glycine	Arginine	Arginine	
Serine	Isolencine	Isoleucine	
Citrulline	Leucine, Valine Histidine, Phenyla- lanine or Pheylalanine + tyrosine		

Part of methionine requirement can be supplied by cystine and that of pheylalinie by tyrosine.

Classification of amino acids on chemical basis

On the basis of chemical composition and chemical properties the 20 amino acids constituting the body tissue can be classified in to the following six groups (Table 3.5).

Table 3.5: Classification of amino acids on chemical basis.

Group	Group name	Amino acids
1.	Aliphatic amino acids	Alanine, glycine, isoleucine, leucine, serine, threonine and valine.
2.	Aromatic amino acids	Phenylalanine and tyrosine
3.	Sulphur amino acids	Cysteine, cystine and methionine
4.	Acidic amino acids	Aspartic acid and glutamic acid
5.	Basic amino acids	Arginine, histidine and lysine
6.	Heterocyclic amino acids	Hydroxyproline, proline and tryptophane

Derived amino acids or special amino acids

Some proteins are formed of derivatives of some normal amino acids or protein. These amino acids are known as derived amino acid or special amino acid. Some examples of derived amino acids encountered in body and participate in specific physiological functions are presented in table. 3.6

Table 3.6: Examples of derived (Special) amino acids.

Protein	Derived amino acid	Mother amino acid
Collagen	Hydroxylysine	Lysine
	Hydroxyproline	Proline
Thyroglobulin	Triidothyronine	Tyrosine
	TetraiodothyronineOr thyroxine	Tyrosine
Thrombin	Gama-carboxyglutamic acid	Glutamic acid
	Gama-aminobutyric acid	Glutamic acid
	Cysteine	Conjugation of two molucules of cystine

Types of tissue proteins in the body

Proteins in different forms present in the body are associated with different kinds of specific roles as shown in Table 3.7.

Table 3.7: Types of tissue proteins in the body

1.	Collagen	Fibrous tissues like ligaments and tendons.
2.	Elastin	Fibrous tissues like ligaments and tendons
3.	Myofibrilar proteions	Proteins of sarcoplasma contains several enzymes required for muscle metabolism, mitochondrial parts and particles of cytoplasmic reticulum.
4.	Contractile proteins	Actin, myosin and tropomyosin B are involved in muscle contraction.
5.	Keratins	Proteins farming feather, beak and claws in poultry.
6.	Blood proteins	Albumin, globulins, thromboplastin, fibrinogen, haemoglobin, apoproteins, enzymes, lipoproteins, protein-harmones, peptides
7.	Enzymes	Almost all enzymes are protein. Large number with specific functsion are present in the body.
8.	Harmones	Harmones are synthesized in glands and transported through circulation to target organs.
9.	Metabolically active peptides and polypeptides'	Insulin like growth factor I and II (IGFI and IGFII), transforming growth factor - beta (GGFB), fibroblast growth factor (FGF), nerve growth factor (NGF)
10.	Immune Antibodies	These are fraction of total body proteins but very important for protection against specific infectious diseases.

Prions

These are the smallest known microorganisms capable of multiplying at very fast rate in animal body. These are considered abnormal or abusive form of a normal cellular protein. Molecular size of prions is about 55 KDa and composed of pathogenic isoform of a prion protein (PrP) which has been termed PrP-Sc. Prion cause many neurological diseases in mammals like scrapie in sheep and mad cow disease. The groups of diseases caused by prions are called transmisible spongiform encephalopathies. Prions have affinity with brain and nervous tissue.

LIPIDS

Lipids are organic components of plants and animals constituted of fat and fat soluble subestances. These are either insoluble or sparingly soluble in water but highly soluble in organic solvents. Some of the lipids are essential for many biochemical reactions and physiological functions of the living body. Fat and oils are the richest sources of dietary energy for the animals including poultry.

Constituents of lipids

The main basic constituents of lipids are the carbon, hydrogen and relatively small amount of oxygen. These constitute the major part of the body lipids. Some of the lipids contain substances other than triglycerides or triacyglycerols. These are compound and derived lipids and contain different kinds of organic and inorganic substances.

Characteristics of lipids

1. Specific density of lipids is ligher than the water due to which they form top layer on mixing with water.
2. Lipids are soluble in organic solvents like ether, hexane and chloroform etc.
3. Lipids are normally insoluble in water except some sparingly soluble derived lipids.
4. These are condensed sources of energy for the animals. Some of the fatty acids are essential for the normal physiological function of the animal body.

Classification of lipids

Lipids are classified into three broad classes of simple lipids, compound lipids and derived lipids.

1. **Simple lipids:** These are esters of fatty acids with glycerol and other alcohols. Simple lipids were earlier classified into fats, oils and waxes but later on fats and oils were merged together being similar in chemical constituents and called fats.

   Fat: These are the esters of fatty acid with glycerol only. In common non scientic terms the triglyceride compounds in solid form at room temperature are fats and other present in fluid forms are oils. However, there are other confusing situations encountered in normal practice like fats of coconut, mahua seed and salseed are solid at room temperature are called oils due to their vegetable origin. Actually the ratios of unsaturated and saturated fatty acids in the triglycerides are responsible for the solid or liquid state at room temperature (20°C). In order to avoid this confusion all pure triglycerides are now called fats.

Waxes: These are also simple lipids and esters of fatty acids with alcohol other than glycerol.

2. **Compound lipids:** These are esters of fatty acids with alcohols which also contain non lipid and proteins. These compound lipids are further named on the basis of non-lipid constituent of the compound like phospholipids, glycolipids, and aminolipids or lipoprotein.

   Phospholipids: The lipids containing phosphorus and nitrogen are phospholipids. Examples are phosphoglycerides or phosphatids. Common phospholipids found in the body are lecithin and cephalin. In lecithin's phosphoric acid combines with the nitrogenous base choline and the term phosphadylcholines is more specific for the compounds. In cephalins phosphoric acid combines with ethanolamine and the compounds are also called phosphatidyl ethanolamines.

   Sphingomyelins are different than the lecithins and cephalins. In place of glycerol the constituent is sphingosine. The terminal hydroxyl group joins with phosphoric acid and not a sugar residue. In sphingomyelines, amino groups link with carboxyl group and long chain fatty acids are formed by linkage with peptide bonds.

3. **Derived lipids:** The lipid fractions and other compounds produced from the hydrolysis of simple and compound lipids are know as derived lipids.

Pseudolipids or lipids like substances

These are generally substances extracted with lipids by organic solvents but these are not fats. Such substances are pigments, strols, terpenes and prostaglandins.

Saturated and unsaturated fatty acids

The fatty acids without double bond in molecular structure are known as saturated fatty acids. Unsaturated fatty acids contain one or more number of double bonds in their molecular structure and the degree of unsaturation is determined by the number of double bonds in the molecular structure.

Omega-3 and omega-6 fatty acids

The polyunsaturated fatty acids (PUFA) containing 3 and 6 number of double bonds in their molecular structure are known as omega-3 and omega-6 fatty acids respectively. Linolenic (18 carbons and 3 double bonds) is an omega 3 fatty acid and docosahexaenoic acid (22 carbons and 6 double bonds) is omega-6 fatty acids. These are found plenty in fish oil and some of the vegetable oils. These are protective substances for heart and in recent years have acquired important place in nutrition and dietary managements.

Essential fatty acids

The polyunsaturated fatty asids essential for the normal physiological functions in poultry birds and not synthesized in the body are known as essential fatty acids. The essential fatty acids to be supplied in the diets of poultry are linoleic acid, gama- linoleic acid and arachidonic acid. In true sense linoleic acid is the only essential fatty acid required to be supplied in the diet and it is converted into arachidonic acid from the linoleic acid ingested in the diet.

Stability of dietary fats and Rancidity

Some natural antioxidants present in the fats provide protection against oxidation. The concentration of natural antioxidants is variable and exhaustion of antioxidants causes oxidation leading to rancidity. Oxidation is more rapid in unsaturated fatty acids and carbon atoms adjacent to double bonds are highly susceptible. The process of oxidation results in the production of short chain compounds and liberation of free radical. Free radicals are highly reactive and enhance the rate of oxidation. The oxidation of saturated fatty acids results in the production of ketones which are responsible for the development of unpleasant odour and decreased palatability and nutritive value. The process is called rancidity.

Functions of lipids

1. Minimum true fat is essential for the absorption of fat soluble vitamins in the body.
2. Fat is also essential for the absorption of carotenoids necessary for the supply of vitamin A activity and imparting yellow colour to egg yolk.
3. Phospholipids in lipoproteins complex are essential for the functions of cell membranes.
4. Glycolipids are the components of brain and nervous system.
5. Adipose tissues are the store houses of energy in condensed form for energy supply. Depot fats are metabolized for blood glucose supply by gluconeogenesis.
6. Dietary fat reduces heat increment in the body.

7. Cholesterol is essential for the normal functions of cells. Production of 7- dehydrocholesterol is essential for the synthesis of active form of vitamin D, i.e., cholecalciferol.
8. Cholesterol is utilized for the synthesis of bile acids necessary for fat digestion and absorption.
9. Essential fatty acids are important for the maintenance of skin health and function of cell membranes and play role in immune modulation.
10. Maintenance of normal reproductive functions like spermatogenesis in male bird and hatchability of eggs in female birds.

ENERGY

Energy is the fuel for the normal physiological functions of the body. In biological terms energy is the chemical constituents of the body yielding calorie through biological oxidation for the vital functions. The heat liberated from the oxidation of glucose is the ultimate form necessary for body functions.

Sources of energy in the body

1. Glucose in the body fluid specially the blood glucose.
2. Glycogen (animal starch) present in the muscles and liver.
3. Adipose tissue in the body.
4. Muscles of the body.

Sources of dietary energy

The sources of dietary energy are carbohydrates, fats and proteins. These are available from food grains, tubers, roots, fats and animal tissues (muscles and fatty tissues). The foods are digested, metabolized and utilized for the maintenance, growth and egg production in poultry.

Units of energy used in poultry nutrition

Generally big calorie or kilocalorie (kcal) or mega calorie (mcal) and joule, kilo joule (kj) and mega joule (mj) are used in different countries for the feeding of poultry birds. A small calorie is the amount of heat required to increase the temperature of 1 g water from 14.5° to 15.5°C. The heat generated from the foods is measured with the help of a bomb calorimeter. Joule is smaller unit than the caloric and in nutrition including poultry nutrition kilocalorie (kcal) or, kilo joule (kj) is mostly used for the expression of energy. For the conversion of calorie to joule the calorie value is multiplied by 4.1854 and joule is divided by 4.1854 to get calorie.

Functions of energy

1. Maintenance of life through the maintenance of normal vital functions of the body.
2. Utilization of nutrients for growth production.
3. Energy supply for metabolism.
4. Energy supply for essential physical activities.
5. Optimum reproductive performances.
6. To provide protection.

Storage of energy

In addition to normal production of skeleton, muscle, skin and feather etc, the excess intake of energy is stored in the chemical form of fat in adipose tissue.

Utilization of stored energy in the body

In case of energy deficiency caused by starvation or other stress the fat stored in the body is mobilized and utilized for the production of glucose via gluconeogenesis. The glucose, thus produced is metabolized for the supply of ATP, the functional form of biological energy participating in physiological functions.

Gross energy (GE)

The total amount of heat produced from complete oxidation or burning of organic matter of a food or other organic substance is known as gross energy. It is also called heat of combustion.

Metabolizable energy (ME)

Since faeces and urine are voided together in the birds, it is more useful and easy to determine metabolizable energy (ME) for practical feeding. Metabolizable energy is the part of gross energy intake minus the gross energy voided in excreta.

Digestible energy (DE)

A study of digestible energy is a subject of academic interest and has little practical utility in poultry production. Sturkie made experimental models with the help of surgery. The ureters were separated from the urodium of the cloaca and joined near the vent for independent excretion of urine. However, it is very expansive and labour intensive to maintain such models.

The other method is on the basis of assumption that grayish white cap on the poultry excreta is urine and faeces can be obtained by its separation. This method is also not free from error.

MINERALS

These are main components of structural frame work of the body besides playing important roles in various physiological functions as constituents, cofactors and activators.

The list of nutritionally essential elements has increased from 13 up to 1960 to mere than 20 by this time. However, there is no change in the list of macro or major 7 elements, calcium (Ca), phosphorus (P), magnesium (Mg), sodium (Na), potassium (K), chlorine (Cl) and sulphur (S) but the list of micro minerals or trace minerals is increasing and exceeded 13 from the earlier 6 trace elements comprising of iron, iodine, copper, cobalt, manganese and zinc. It is expected that some more trace elements may be found involved in some physiological functions in the animal body. The present status of micro minerals is presented in table.3.8.

Table 3.8: Micro (trace) minerals of nutritional value

Trace minerals		Probable trace minerals	
Boron	B	Aluminium	Al
Cobalt	Co	Arsenic	As
Chromium	Cr	Bromine	Br
Copper	Cu	Lead	Pb
Fluorine	F	Lithium	Li
Iodine	I	Nickel	Ni
Iron	Fe	Tin	Sn
Manganese	Mn	Vanadium	V
Molybdenum	Mo		
Selenium	Se		
Silicon	Si		
Zinc	Zn		

The dietary mineral elements requirement higher than 0.1 per cent in the dry matter of feed is known as major or macro minerals and those present below 0.1 per cent are micro or trace minerals. The concentration of trace elements in animal body is generally below 50 mg per 100 g of dry matter.

Unwanted minerals in the body

Several unwanted minerals may be found in the body and their presence depends on the presence in soil and water of the area and the up take of the elements by the food crops cultivated on such soils followed by absorption from food in the body.

Toxic levels of minerals

Many minerals of dietary and nutritional importance in traces become harmful and toxic in higher concentration present in food or drinking water. Higher intake of selenium, fluorine and molybdenum etc is highly toxic.

Importance of balance among the minerals intake

The availability and physiological functions of many minerals are interdependent and absorption and utilization ratio between Ca; P should be 1:1 to 2:1 in the diets of growing birds but in the diets of laying hens a much wide ratio of 6:1 or even higher is required to cope with the utilization of calcium for eggshell synthesis. Balance among dietary mineral is necessary for proper metabolism. Higher concentration of calcium in the diet increases the requirement of phosphorus. Imbalances in the dietary anions and cations adversely affect acid-base balances which are often reflected in lower body growth and decreased egg production. It is, therefore, essential to provide balanced amount of available minerals.

Bioavailability of minerals from feeds and supplements

Significant effect of feed composition occurs on the bioavailability of minerals from the feeds and sources of mineral supplements. The phytates, oxalates and some other complex forms are sparingly available to poultry. Similarly bioavailability of minerals from the inorganic sources is also variable, e.g. bioavailability of phosphorus from dicalcium phosphate is normally higher than the phosphorus in unprocessed rock phosphate. For increasing the bioavailability of phosphorus and other minerals present in phytate forms phytase enzyme is supplemented in the poultry diets.

It is often claimed that bioavailability of chellated organic minerals is higher than the inorganic salts. This effect can not be generalized and merits and economic implications of the use of organic forms of minerals as poultry feed supplements should be given due consideration.

Minerals ingested in the diets are changed in to three forms during the course of digestion and normally found in the digesta. These are ionic form, soluble metallo-organic complex compounds or insoluble compounds. The bioavailability of ionic form is most efficient followed by the soluble metallo-organic compounds including the chelates. Inorganic complex compounds are almost entirely voided in the excreta.

Bioavailability of minerals is also influenced by the age, stage of production and deficiency. Bioavailability of all essential minerals is high in young growing animals, while that of calcium is high in laying birds. The absorption of calcium and phosphorus increases during deficiencies like rickets and osteoporosis and that of iron during anaemia. Some examples of bioavailability of trace minerals from different kinds of inorganic and organic mineral supplements are compiled in Table 3.9.

Table 3.9: Bioavailability of trace minerals from some inorganic and organic supplements.

Sources	Trace element	Bioavailability %
A. Inorganic mineral supplements		
Copper carbonate	Copper	65
Copper sulphate	Copper	100
Ferrous chloride	Iron	100
Ferric oxide	Iron	10 (negligible, not used)
Ferrous sulphate	Iron	100
Manganese carbonate	Manganese	55
Manganese sulphate	Manganese	100
Sodium selenite	Selenium	100
Zinc chloride	Zinc	100
Zinc sulphate	Zinc	100
B. Organic mineral supplements		
Copper lysine	Copper	100
Copper methionine	Copper	90
Ferric citrate	Iron	75
Manganese-methionine	Manganese	100
Selenium cystine	Selenium	100
Selenium methionine	Selenium	100
Zinc methionine	Zinc	100

Note: use of organic mineral supplements in place of inorganic salts with bioavailability more than 70 per cent may be considered undesirable because it will increase feeding cost without considerable impact on health and productivity.

Sources of minerals supplements for poultry: The demand of minerals is very high for the domesticated birds due to many fold increase in their production when compared to wild forms. This change in production and farming (Rearing) system requires supplementation of dietary essential minerals. In practical poultry production calcium, phosphorus and trace minerals receive special attention. Content of minerals are different in different sources. Hence, it is important to know the content of minerals in different sources (Table 3.10 and 3.11) for the selection at the time of procurement for compounded feed preparation.

Table 3.10: Common sources of calcium and phosphorus supplements.

Source	Chemical formula	Ca%	P%
Steamed bone meal	Complex	29	12
Calcium carbonate	$CaCO_3$	38	–
Marble powder		35-38	–
Dicalcium phosphate	$CaH PO_4 . H_2O$	22	18
Monocalcium phosphate	$Ca(H_2PO_4)_2$	16	21
Oester shell grit	–	38	–
Lime stone	$CaCO_3$	35-38	
Calcium sulphate	$CaSO_4.2H_2O$	22	
Deflorinated Rock phosphate		32	18
Soft Rock phosphate		17	9

Table 3.11: Common sources of trace minerals.

Salt name	Chemical formula	Mineral	Content (%)
Blue vitrial cr Copper Sulphates			
Copper sulphate (Monohydrate)	$CuSO_4.H_2O$	Copper	
Copper sulphate(Pentahydrate)	$CuSO_4.5H_2O$	Copper	
Cupric carbonate	$CuCO_3$	Copper	53
Cupric oxide	CuO	Copper	75
Ferrous carbonate	$FeCO_3$	Iron	43
Ferrous sulphate(green Vitrial)	$FeSO_4.7H_2O$	Iron	21
Manganese carbonate	$MnCO_3$	Manganese	47
Magnous sulphateMono hydrate	$MnSO_4.H_2O$	Manganese	25
Magnous sulphate Pentahydrate	$MnSO_4.5H_2O$	Manganese	22

[Table Contd.

Contd. Table]

Salt name	Chemical formula	Mineral	Content (%)
Potassium iodate	KIO_3	Iodine	59
Potassium iodide	KI	Iodine	76
Sodium selanate	Na_2SeO_4	Selenium	41
Sodium selenite	Na_2SeO_3	Selenium	45
Zinc oxide	ZnO	Zinc	73
Zinc sulphate	$ZnSO_4$	Zinc	22

Minerals from animal sources like bone meal and bone-cum-meat meal are avoided in the diets of animals.

VITAMINS

These are essential micro nutrients of organic composition but mostly unrelated chemical nature. These are dietary requirements in traces and deficiency results in occurrence of health problems and conspicuous drop in production. The terminology "Vitamin" was adopted due to its historical importance. The name vitamin was given by Funk(1912) to certain minute organic constituents of foods capable of preventing diseases like beri-beri, pellagra etc. Exact date of discovery is not yet known. As early as 1753 a British doctor of navy named Lind observed that inclusion of juicy fruits and salads in the diets prevented the occurrence of scurvy. Studies of ancient literatures and the literatures of Vedic and pre-historic period of India may unveil many important informations on the role of micronutrients like vitamins.

Vitamins are classified into two main groups of fat soluble vitamins and the water soluble vitamins.

Fat soluble vitamins: These are soluble in fat and the organic solvents dissolving the fats. The precursors of vitamins are also soluble in fat and organic solvents. The fat soluble vitamins may be present in more than one chemical form in the feed sources but active form useful in the body is mostly one. However, most of the other forms ingested in feeds are converted into biological active form in the digestive system. The members of fat soluble vitamins are A, D, E and K.

Water soluble vitamins: These are B-complex vitamins and vitamin C. All are soluble in water and mostly highly unstable. Regular dietary supplementation is required. Most of the members of vitamin B complex group are coenzymes or prosthetic group required at different stages in the metabolism of various nutrients and energy utilization for body vital functions. The members of vitamin B –

complex group are thiamine (B_1), riboflavine (B_2), biotin, nicotinamide, pyridoxine (Vitamin B6), folic acid, pantothenic acid, choline, vitamin B12 and vitamin C or Ascorbic acid.

Sources of vitamins

Vitamins are available to poultry both from the plant and animal sources, either in the active forms or in the form of precursors which are transformed into active form in the gastro-intestinal tract.

ENZYMES

Enzymes are not considered as nutrients but these are essential for the availability of nutrients for the normal functions of body. These are compounds required to act as catalyst for the storage as well as release of the energy as per the physiological requirements in the body. Various chemical reactions are necessary for the change of energy from storage form to functional form and vice versa. On the basis of functions the enzymes may be grouped in to the following categories.

A. **The oxido-reductases:** These are actively involved in the transfer of hydrogen, oxygen or electrons from one molecule to another in the body. Following are the various groups of enzymes of oxido-reductase category.

Catalases
Dehydrogenases
Hydroxylases
Oxygenases
Oxidases and
Peroxidases

B. **The transferases:** These are enzymes responsible for the transfer of functional groups like acetyl group, amino group and phosphate group from one molecule to another. The various transferases are

Acyl transferase
Glucosyl transferase
Phosphoryl transferase
Transaldolases
Transketolases
Kinases
Phosphomutases

C. **The Hydrolases:** The enzymes catalyzing the hydrolytic division of molecules in the body are hydrolases. The hydrolytic enzymes are specific for the cleavage of fat molecules and protein molecules. Various type of hydrolases are:

Amidases
Deaminases
Phospholipases
Phosphatases

Esterases
Glycosidases
Peptidases
Ribonucleases
Thiolases

D. The lyases: These enzymes are responsible for non-hydrolytic cleavage and removal of certain groups from the molecule like removal of carbon dioxide (CO_2) in decarboxylation reaction and amino group (-NH_2) in deamination reaction. Some of the enzymes of this category are:

Aldolases
Dehydrolases
Hydralases
Lyases
Synthases

E. The Isomerases: These are enzymes responsible for catalyzing reactions for the rearrangements of position of elements or groups within the molecule. The various groups of enzymes are:

Isomerases
Racemases
Mutases (not all mutases)

F. The Ligases: The enzymes catalyzing the formation of new molecules from two molecules are the ligases. The reaction receives energy from the break up of ATP to AMP in case of production of Acetyl-CoA from the reaction of acetic acid and co enzyme A.

4

FUNCTIONS OF NUTRIENTS

Nutrients are the compounds necessary for the maintenance of life. Greater part of a living creature is made of water followed by organic and inorganic compounds. Organic nutrients constitute the largest proportion of poultry body, the highest being protein followed by lipids, minerals and vitamins. The functions of feed and water are the result of complex biochemical activities in the body for the utilization of ingested foods for the maintenance of all physiological functions of the body. The main physiological functions of a living animal may be listed as follows:

Growth

1. **Maintenance of life with the help of blood and blood circulation for**
   - (i) Gaseous exchange, i.e. supply of oxygen to each cell of the body and collection of carbon dioxide from each cell for expulsion out of the body by expiration.
   - (ii) Transportation of nutrients to each organ and cells of the body.
   - (iii) Maintenance of acid-base balance, osmotic pressure and homoestasis.
2. **Reproduction:** This includes birth of off springs and their maintenance till they become capable of surviving on non-maternal foods like milk in mammals and eggs in aves, reptiles and pisces.

All these functions in living body are accomplished by reconstitution of dietary nutrients in the body in different organs, particularly the glands (liver, pancreas, spleen, kidneys) and tissues.

1. WATER

Water is the most important constituent of body and survival is not possible without water. More than two-third of animal body is water and its proportion in youngs is more than 70 percent of body mass.

Functions of water in the body

1. Maintenance of equilibrium among the fluid compartments (blood, lymph, interstitial fluid and intracellular fluid). The function is known as homoeostasis.
2. It is an essential medium for the optimum digestion because effective action of enzymes needs adequate water.
3. Exchange of nutrients and metabolites upto cellular level.
4. Lubricant.
5. Constituent of products, viz. chicken egg contains about 67% water and meat contains 75% water in broilers but less (about 70%) in culled birds.
6. Maintenance of body temperature.

Symptoms of deficiency

1. Loss of appetite resulting in lowered feed intake; complete cessation causing decreased growth and egg production.
2. Dehydration and ruffled feathers.
3. Death on loss of more than 20 percent water from the body in most of the species.

2. CARBOHYDRATES

1. Staple food constituting more than 50 percent of daily diets.
2. Main source of metabolizable energy and productive energy.

Functions

1. Energy supply for life and production.
2. Feeling of satiety.
3. Maintenance of blood sugar level.
4. Soluble non starch polysaccharides (NSPs) like pectins, arabinoxylans, beta-glucons, xylans and hemicellulose are indigestible and also interfere in the digestion of other nutrients. However, these absorb water and help smooth passage of ingesta and digesta.
5. Storage of energy in the form of glycogen in liver and muscles and also considerable increase in blood glucose (130-270 mg/dl) level.

6. Formation of ribose essential for AMP, ADP, ATP, RNA and DNA.
7. Essential for metabolism of protein and fat. In case of hypoglycemia fat metabolism is incomplete and ketone bodies accumulate.

3. PROTEINS AND AMINO ACIDS

Proteins are the polymers of essential and non-essential amino acids required for the formation of body tissues. There are also free circulating proteins particularly conjugated proteins and other responsible for the management of vital functions in the body. The properties and functions of proteins depend on the length of peptide chain, structural arrangements of the amino acids and incorporation of other nutrients or elements in the protein structure (Conjugated proteins like lipoproteins, glycoproteins and metaloproteins etc.).

Functions of proteins

1. Formation of cells and tissues including gametogenesis.
2. Connective tissues like ligaments and tendons composed of fibrous proteins (collagen and elastin) responsible for movement and also the formation of skin and feathers.
3. Glycoproteins and lipoproteins are the constituents of cell membrane and responsible for the maintenance of homoeostasis and osmotic pressure by controlled movements of fluids, nutrients and metabolites across the membranes.
4. Blood proteins fibrin and thrombin with other factors are responsible for blood clotting.
5. Gaseous exchange, haemoglobin in blood and myoglobin in muscles are responsible for transportation of oxygen upto cells and removal of carbon dioxide from cells to lungs for expulsion in expired gas.
6. Transportation of nutrients to target organs for growth, repair, replacement and storage etc.
7. Transmission of genetic materials in gametes (eggs and sperms) and nutrients supply for gametes.
8. All enzymes and most of the hormones are proteins. These are essential for metabolism and vital functions.

9. Antibodies providing protection against diseases are proteins.
10. Provides metabolizable energy by gluconeogenesis in deficit glucose supply from deficit intake of carbohydrates and fats.

Deficiency symptoms

1. Loss of body weight.
2. Lowered growth and egg production.
3. Increased susceptibility to infection due to immunosuppression.
4. Decreased feed conversion efficiency.
5. Protein deficiency with high energy intake from carbohydrates and fats causes excessive fat deposition (obesity).
6. Deficiency of specific amino acid like arginine causes slow feather growth in chicks and ruffled feather, cannibalism and feather pecking in all stages.
7. Deficiency of methionine and choline causes perosis or slipped tendon. Manganese deficiency is also responsible for perosis.

Non Protein nitrogenous compounds (Purine and pyrimidine) in the body

These are nucleotides present in the cells of all animals and associated with energy metabolism (energy transfer). These are supplied in feeds and also “de novo” synthesized in the body. The bases in purine are adenine and guanine, and that in pyrimidine are cytosine, thymine and uracil. The condensation products of any of these bases is called nucleoside, which becomes nucleotide by phosphorylation. The nucleic acids, viz. ribonucleic acid (RNA) and deoxyribonucleic acid (DNA) are the polymerization products of nucleic acids. The nucleotides are also utilized for the synthesis of co-enzymes, nicotinamide adenosine dinucleotide (NAD) and flavine adenosine dinucleotide (FAD) required for the transfer of energy at cellular level.

Deficiency symptoms

Deficiency is produced indirectly due to deficit supply of nicotinamide and flavine. The symptoms are unthriftyness and failure of many vital functions in the body. Failure of intermediary metabolism produces complex disorders in the body functions.

4. FATS

The proportion of fats is highly variable and depends on the species, breed, age, sex, level of nutrition, composisiton of diets and production level (Egg production). For normal physiological functions the level of body fat in birds should be more than 2 percent. Body fat content in non producing obese bird may be upto 50 per cent, although such high level is a rare incidence.

Functions

1. Condensed source of energy. About 2.25 times more than the available energy from carbohydrates or proteins.
2. Carrier of fat soluble vitamins and also the precursors of vitamin like carotenoids that impart yellow colour of egg yolk. It shows that all dietary carotenoids ingested in feeds are not converted to vitamin A activity (Retinol) in intestinal mucosa and liver parenchymal cells.
3. Structural and functional components of cell membranes
4. Storage of energy in fat depots.

Essential fatty acids

Although three fatty acids, i.e., linoleic, linolenic and arachidonic acid have been identified as essential fatty acids for the animals but for the birds only linoleic acid is strictly essential. Vegetable oils are rich sources of linoleic acid.

Unsaturated fatty acids

The fatty acids containing double bond (s) in their structure are known as unsaturated fatty acids. A fatty acid with single double bond in molecule is known as mono-unsaturated fatty acid (MUFA) and those with two or more double bonds in the molecule are polyunsaturated fatty acids (PUFA). These are considered better than the saturated fatty acids.

Blood lipids

The sources of blood or plasma lipids are:

1. Absorbed from the digested feeds.
2. Lipids synthesized in the liver.
3. Lipids mobilized from fat depots in the body.

The plasma lipids are found in the following chemical forms:

1. Chylomicrons
2. Lipoproteins: These are very low density lipoproteins (VLDL), low density lipo-proteins (LDL) and high density lipo-protein (HDL).

Symptoms of deficiency

1. Poor growth
2. Reduction in production and size of eggs.
3. Reduced fertility in both sexes
4. Poor hatchability

Symptoms of excess fat intake

1. Development of obesity
2. Occurrence of cardio-vascular disorders.

   VLDL and LDL are claimed to be more harmful

5. ENERGY

Energy is the heat generated from the metabolism of organic nutrients-carbohydrates, lipids and proteins. Survival is not possible without energy supply as all vital functions need energy.

Symptoms of deficiency

1. Unthriftness
2. Loss of body weight.
3. Reduction in metabolism and other body functions.
4. Fall in egg production.
5. Deformity in phenotypic appearance.

Symptoms of high excess

1. Development of obesity.
2. Occurrence of other deficiency disorders caused by the imbalance of nutrients ingested.

6. MINERALS

General functions of minerals.

1. Development and maintenance of skeletal system.
2. Maintenance of acid-base balance by maintaining the ratios of anions (chloride, iodide, phosphates) and cations (Calcium, Magnesium, Potassium, Sodium, Iron, Mangenese and Zinc).
3. Cellular respiration with the help of haemoglobin and myoglobin.
4. Immunomodulator.
5. Catalyst/cofactor in enzyme and hormones activities.

Minerals constitute about 4 percent dry matter in the body. All dietary minerals involved in various physiological functions are exogenous and essential because these can not be produced in the body. Seven minerals constituting more than 0.1 per cent are called major or macro minerals and include calcium, phosphorus, magnesium, sodium, potassium, chloride and sulphur.

The minerals requirement less than 0.1 per cent in the body are known as micro minerals or trace elements. The nutritional relationship among the minerals is very important because it is synergistic as well as antagonistic among different dietary essential minerals. Likewise there are certain minerals that are nutritionally essential in traces but turn harmful or even toxic on increasing the concentration beyond the maximum tolerance level. Therefore, it is essential to maintain concentration as well as ratio among the nutritional minerals.

Calcium and phosphorus

In the animal body calcium (Ca) and phosphorus (P) are required in the ratio of 1:1 to 2:1 and vitamin D3 for optimum utilization.

1. Largest quantity of calcium and phosphorus are used for skeletal development.
2. Common feeds (grains, oilseed etc.) of plant origin are deficient in calcium, phosphorus and many other minerals and required supplementation of richer sources.
3. For optimum utilization dietary ratio of Ca:P should range from 1;1 to 2:1 preferably 2:1, but in feeds of laying hens this ratio changes to about 5:1 due to continuous turn over of the large quantity of calcium for the formation of egg shell.

4. Apparently the quantity of P in feeds of plant sources may be considered reasonably adequate for adult birds but it is not available due to presence in phytate form. Liberation of phytate-P (PP) requires the enzyme phytase which is not produced in the birds.
5. The phosphorus found in the feeds of animal sources like bone meal, bone ash, fish meal and meat-cum bone meal are non-phytic phosphorus (NPP) which are readily available to birds and other animals.
6. Supplementation of not only Ca and P but many other essential minerals is necessary for the balancing of diets.

Functions of Calcium (Ca)

1. Skeletal development and maintenance (bone contains almost 99% Ca) in all birds.
2. Egg shell formation in laying birds. Egg shell contains 3-4% calcium.
3. Maintenance of egg shell permeability.
4. Neuro-muscular excitability depends on Ca level.
5. Active role in blood clotting.
6. Co-factor in enzyme systems.

Functions of phosphorus (P)

1. Bone formation, almost 80 percent absorbed P is used for bone formation.
2. Functional constituents, viz. phosphoproteins, phospholipids and nucleoproteins. Phosphorus is essential for the functions of cells.
3. Essential for energy metabolism particularly the release of energy for body functions and also storage of excess energy as body reserve.
4. Cell formation and repair of wear and tear.

Deficiency symptoms of Ca and P

1. Decreased feed intake due to inappetence.
2. Loss of body weight and weakness.
3. Rickets in growing chicks.

4. Lowered egg production.
5. Reduction in egg weight.
6. Thin shell formation.
7. Shellless egg laying in extreme cases.
8. Mottled egg yolk.
9. Decreased hatchability in breeding flock.
10. Weak chicks are hatched. Post hatching chick mortality is high and performance of survivers is poor.

Sodium or Natrium (Na) Chloride

Due to more similarity in biological functions, utilization and requirement of the sodium (Na) and chloride (Cl) are considered together as sodium chloride for the nutritional purpose. Common salt or sodium chloride is an important component of diets of animals including birds.

Common poultry feeds of plant origin like food grains, leguminous seed and oilseed cakes are deficient in sodium chloride. But, the feeds of animal origin like fish meal, lobster meal, meat meal and blood meal are good sources of sodium and chloride. Most of the fish meals often contain very high salt content.

Functions

Greater proportion of sodium is present in the body fluids and soft tissues and associated with the following important functions in the body.

1. Maintenance of acid-base balance reflected as pH and osmotic functions of the body fluids.
2. Transmission of nerve impulse.
3. Absorption of carbohydrates (sugars) and amino acids from the alimentary canal.
4. Essential constituent of gastric juice
5. Enteric amylase is activated by salt.
6. Role in proper functioning of many enzymes of nucleus and mitochondria of the cells.

Deficiency symptoms

1. Loss of appetite proportional to salt deficiency is most common sign of deficiency.
2. This is followed by decreased feed intake, lowered digestibility and reduced nutrients availability for body functions.
3. Growth retardation in young and unthriftiness in others are the result of mal function of energy-protein metabolism.
4. Impaired gonadal function adversely affecting oogenesis and spermatogenesis.
5. Fall in egg production, fertility and hatchability of eggs.
6. Occurrence of cannibalism.
7. Enhances moulting in laying hens.

Salt toxicity

Many times poultry feed compounding companies inadvertently ignore the salt content in fish meal and add recommended quantity of common salt. Some time salt containing mineral mixtures may be the cause of salt toxicity. Shifting of birds from other parts to coastal areas where drinking saline water may be the cause of salt toxicity.

Symptoms of salt toxicity

1. Increased craving for drinking water.
2. Occurrence of diarrhoea.
3. Dehydration
4. Decreased body functions.
5. Death

Potassium or Kalium (K)

1. Osmotic regulation of body fluid alongwith sodium chloride, bicarbonates etc.
2. Maintenance of acid-base balance and pH of body fluids.
3. Transportation of nerve impulses to muscles
4. Gaseous exchanges, i.e. transportation of oxygen as oxy-haemoglobin to cells and removal of carbon dioxide as met haemoglobin from the cells.
5. Co-factor in many enzyme systems.

Deficiency symptoms

1. Like salt deficiency loss of appetite accompanied with decreased nutrients supply resulting in lowered growth, production and health status.
2. Occurrence of intra cellular acidosis.
3. Titanic seizure and paralysis.
4. Lowered egg shell quality.

Magnesium (Mg)

Common poultry feeds provide adequate magnesium necessary for normal body functions. Deficiency, if ever recorded may be due to clinico-pathological reasons.

Functions

1. Involved in bone formation.
2. Co-factor in large number of enzyme systems (Thiamine pyrophosphate).
3. Activator for many enzyme systems of metabolic processes (galactokinase, glucokinase, isocitrate dehydrogenase and enolase).
4. Necessary for oxidative phosphorylation in energy metabolism.

Deficiency symptoms

No deficiency occurs in normal feeding conditions. Some experimental and health associated deficiency symptoms are:

1. Inappetance resulting in lowered feed and nutrients intake causing retarded growth, low egg production and weakness.
2. Poor and ruffled feathering.
3. Panting and gasping.
4. Decreased egg size, production and deformed egg shell.

Sulphur (S)

There is no independent role of inorganic sulphur and its inorganic salts in poultry nutrition. The dietary essential organic compounds are the sulphur –amino acids, cystine and methionine. Thus, inorganic sulphur in any form has no role in poultry nutrition. However, sulphur interferes in the normal functions of copper.

7. MINOR OR TRACE MINERALS

Iron, copper, manganese, zinc, iodine, selenium, fluorine and cobalt receive more attention in ration formulation of poultry.

Iron or ferrous (Fe)

Iron is the basis of life maintenance responsible for cellular respiration or gaseous exchange essential for vital functions.

Functions

1. Gaseous exchange at cellular level through haemoglobin in blood and myoglobin in muscles.
2. Component of enzymes like oxidases, oxygenase, peroxidase and catalase.
3. Cytochromes are responsible for electron transport.
4. In coloured birds iron salts are essential for the characteristic pigmentation of feathers and probably also of egg shells. Lysine and folate are associated with iron in development of pigments of coloured feathers.

Conditions affecting iron requirement

1. Dietary ingredients supplying gossypol (in cotton seed meal), phytate in all feeds of plant origin and tannins in some of the feed ingredients like red sorghum (milo), salseed meal, mango seek kernel cake and tamerind seed etc.
2. Excessive blood loss due to dysentery (generally encountered in coccidiosis) and bleeding due to injury.
3. Iron absorption is also reduced due to excess of phosphorus, copper or manganese in the feeds.

Deficiency symptoms

1. Unthriftiness and decreased production.
2. Flushed visible mucous membranes.
3. Macrocytic and hypochromic anaemia.
4. Poor and dull feathers appearance.
5. Various level of depigmentation of feathers in coloured birds.

6. Increased embryonic mortality.
7. Hatched chicks are weak and anaemic resulting in high post hatching chick loss.

Copper or cuprum (Cu)

Copper is essential activator and constituent of important functional components. Availability of copper is significantly reduced by the excess of calcium, molybdenum, iron and sulphur.

Functions

1. Essential for haemoglobin synthesis as a catalyst or co-enzyme.
2. Essential for the utilization of iron for haemoglobin and myoglobin synthesis.
3. It is a constituent of cytochorome oxidase and associated with synthesis of phospholipids which are essential component of myelin sheath.
4. Cytochrome oxidase is also associated with electron transportation system.
5. It is a component of lysyl oxidase involved in cross linking of connective tissue proteins-collagen and elastin.
6. Maintenance of immune status in the body with the involvement of copper, manganese and zinc dependent super dismutase.
7. Play role in osteogenesis.
8. Synthesis of pigments in coloured feather birds.

Deficiency symptoms

1. Anaemia and weakness.
2. Fracture of long bones and lameness.
3. Deformed egg shells and shell less egg laying.
4. Fall in egg production, reproduction and hatchability.
5. Early embryonic mortality (at 3-4 days of incubation).
6. Post mortem reveals enlargement of aorta due to defective elastin formation resulting in thickening and enlargement of aorta. Some times rupture of aorta may be observed.

Manganese (Mn)

Bran, husk, hulls and leguminous leaf meals are satisfactory sources of manganese for poultry. However, it is supplemented in trace to ensure supply for physiological functions.

Functions

1. It is associated with improvement of reproduction and hatchability.
2. Required for growth, egg production and egg shell formation.
3. Mn is constituent of some enzymes like manganese superoxide dismutase, pyruvate carboxylase and arginase.
4. It is a component of mucopolysaccharide and glycoproteins used for the formation of bone matrix.
5. It participates in carbohydrate metabolism.
6. Mn alone and in combination with Mg activates many enzyme system like glycosyl transferase, hydrolases, decarboxylases and kinases.
7. Higher concentration in mitochondria is considered its involvement in partial oxidative phosphorylation.
8. Inhibition of peroxidation of fat.
9. Required for the synthesis of insulin in pancreas.
10. Mn is also involved in the maintenance of body immune system.

Deficiency symptoms

Manganese availability and utilization is adversely affected by the higher concentration of many minerals like calcium, phosphorus, iron and cobalt. The common symptoms of Mn deficiency is more pronounced in chicks and growing birds:

1. Perosis or slipped tendon is most prominent symptom in chicks. However, it should be differentiated from choline deficiency induced perosis. The hock joints are enlarged. Long bones are thickened and shortened. Twisting of tibia and tarso-metatarsal bones. Deformity of condyles of hock joints results in slipped tendon.
2. Reduced reproduction and hatachability. Chondrodystrophy is common in embryonic stage.

3. Bones are weak in newly hatched chicks and other deformities are oedema, bulging abdomen, parrot beak and deformed head. Incoordination of gait and ataxia are observed in surviving chicks.
4. Stargazing or posterior retraction of head in chicks is considered another characteristic symptom of Mn deficiency.
5. Micromelia or shortening of long bones with incomplete and deformed extremities of wing bones, leg bone and spinal column.
6. Loss of libido and aspermia in roosters.

Zinc (Zn)

Zinc is a dietary essential trace mineral and integrated part of several metalloenzymes. It is also activator of many other enzymes. Cereal grains and oil seed cakes are generally deficient in zinc.

Functions

1. Zinc is component of many metalloenzymes like carbonic anhydrase, carboxypeptidase A and B, alkaline phosphatase, DNA polymerase, ribonuclease and many dehydrogenases.
2. It is activator for some of the enzymes.
3. It is associated with the roles of some hormones like insulin, testosterone and steroids.
4. Maintenance of immune system.
5. Required for protein synthesis.
6. Maintenance of water level (homoeostasis) in the body.
7. Antioxidant and protection of cell membranes.

Deficiency symptoms

1. Weak chicks and high early mortality.
2. Enlarged hocks, shortened legs and growth retardation.
3. Lowered weight of thymus, spleen and bursa fabricius.
4. Rough feather and scaly skin.

5. Reduced hatchability and deformed embryonic development, viz. fusion of lumbar vertebrae, missing toe and shortened legs etc.
6. Hatched chicks are weak and unable to stand and walk. Breathing difficulties and death of most chicks due to various degree of starvation.
7. Incomplete and slow feather growth and failure of feather growth in extreme cases.
8. In surviving chicks feed intake is lowered which causes delayed maturity and less egg production.

Iodine

Iodine is a constituent of thyroid hormones thyroxine, triiodothyronine and calcitonin. Iodine deficiency causes compensatory growth in thyroid glands resulting in the formation of goiter.

Functions

1. Constituent of thyroid hormones.
2. Necessary for intermediary metabolism.
3. Maintenance of bone growth.
4. Essential for cell oxidation.
5. Thermoregulation.
6. Reproduction and embryonic development.

Deficiency symptoms

1. Goitre due to abnormal enlargement of thyroid glands. It is an attempt of the gland for maintaining the supply of thyroid hormones.
2. Reduction in egg size and decreased egg production.
3. Retarded growth of chicks.
4. Abnormal feather development.
5. Decreased energy utilization and increased fat deposition.
6. Gonadal degeneration in roosters resulting in reduced sperm production.

Cobalt (Co)

Cobalt is a constituent of vitamin B_{12} and no role of cobalt is known in elemental or any other form. The functions and symptoms of deficiency are similar to that of vitamin B_{12}.

Selenium (Se)

The nutritional role of selenium was recognized in 1957 on observing its preventive effect on exudative diathesis in chicks. Subsequent studies also demonstrated greater similarities between the physiological functions of selenium and vitamin E. Requirement of Se is minute and even little excess intake may be detrimental.

Functions

1. It is a component of enzyme glutathione peroxidase required for the catabolism of peroxides produced during lipid metabolism.
2. Protection of cell membranes from oxidation.
3. It is a constituent of purine and pyrimidine bases responsible for the maintenance of nucleic acids integrity.
4. Maintenance of immunity. In combination with vitamin E it protects leukocytes and macrophages against phagocytosis.
5. Precipitates many heavy metals like cadmium, mercury etc. and significantly reduces absorption, alongwith vitamin E.
6. Maintenance of reproduction.
7. Selenium and vitamin E are responsible for the protection of mitochondria and microsomes responsible for synthesizing antibodies in cells.

Deficiency symptoms

Deficiency symptoms of Se may be independent or associated with vitamin E or vitamin E and with sulphur amino acids also.

1. Pancreatic dystrophy is caused by Se deficiency and cured by Se supplementation. It does not respond to any dose of vitamin E supplementation even for longer duration.
2. Exudative diathesis occurs due to deficiency of either and also responds to treatment with either and combined.

3. Nutritional muscular dystrophy or white muscle disease caused by the deficiency of Se, vitamin E and sulphur amino acids. Degeneration is more prominent in breast muscle. This condition can also be cured by Se or vitamin E supplementation.

Toxic minerals found essential for poultry

Some of the minerals found toxic for poultry are the salts of aluminium, barium, bromine, cadmium, chromium, fluorine, lead, mercury, molybdenum etc.

Molybdenum (Mo)

Molybdenum is essential for optimum growth in poultry but normally it is not a dietary essential because the traces present in the common poultry feeds are adequate to meet the minute requirement.

Silicon (Si)

Silicon has been found essential for normal growth in growing birds but it is not dietary essential. Plenty silicon is found in nature. Fortunately its absorption is extremely low and it is absorbed only in silicic acid form which is available in minute quantity and chicks are saved from harmful effects.

Arsenic (As)

Growth stimulation in chicks has been observed on arsenic supplementation in purified or arsenic free diet. However, it is not dietary essential because food and water contain adequate arsenic. In many parts of India arsenic in water is quite high and had reached toxic level.

Nickel (Ni)

It is an essential trace element for growth, reproduction and functions of liver but it is not a dietary essential due to adequate content in feeds and soil.

Vanadium (V)

It has been found to increase the rate of glucose transportation and possibly metabolism. It is not dietary essential and needs much experimentation for understanding its roles in metabolism.

Suspected essential trace elements

Some more elements like tin (stanus, Sn), barium (Ba), bromine (Br), Cesium (Cs) and rubidium are suspected for playing some role in metabolism. Extensive studies are required before consideration of these elements for their essentiality.

Bioavailability of minerals

Dietary supplementation of minerals does not ensure bioavailability which is affected by several factors like:

1. Interaction with other minerals.
2. Formation of organic complex with phytic acid.
3. Age, production and reproduction stages
4. Environmental factors

Tolerance and toxic levels of dietary essential and some other mineral normally present in feeds, water and environment.

The available information on tolerance, toxic level and characteristic signs of toxicity are summarized in Table 4.1.

Table 4.1: Tolerance and toxicity levels and characteristic toxicity symptoms of some dietary essential elements.

Mineral	Tolerance level	Toxic level	Symptoms of excess intake
Established essential macro minerals			
Calcium	Chicks/broilers 1.2% layers 5%	–	Deficiency of P, Mg, Fe, I, Mn, Zn etc.
Phosphorus	0.8 -1%	–	Almost similar to Ca excess
Magnesium	Chicks 0.3% Others 0.5 %	more thah 1%	Retarded growth, fall in egg production & thin shell. Lower bone mineralization. Soiling of vent, eggs and feather due to wet droppings.
Common salt	1.5%	2%	Reduced growth, increased water intake, loose excreta, fall in egg production.
Potassium		2%	Wet excreta
Essential micro or trace minerals			
Iron	0.1%	0.45%	Reduced availability of vitamins and other minerals. Reduced P availability due to insoluble complex formation. Rickets in chicks. Thin egg shell

[Table Contd.

Contd. Table]

Mineral	Tolerance level	Toxic level	Symptoms of excess intake
Iodine	0.03%	0.05%	Disturbed metabolism. Goitre due to hyperthyroidism. Reduced growth. Reduced egg production and egg size. Reproductive disorders. Lowered hatchability and weak hatching.
Copper	0.03%	0.08%	Anaemia, Erosion of gizzard and necrosis of liver parenchyma. Loss of vitamin E Fall in growth rate
Cobalt	0.001%	0.01%	Loss of appetite. Lowered growth.
Manganese	0.02%	0.4%	Lowered bone density. Reduced egg production thin egg shell.
Zinc	0.1%	0.15%	Exudative diathesis, muscular dystrophy, lowered bone mineralization. Lowered growth.
Selenium	2mg/kg	10mg/kg	Reproductive disorders. Lowered egg production, low hatchability. Deformity of legs and wings, defective eye development or lack of eyes.

Toxic minerals

Almost all minerals produce harmful effects on increasing concentration beyond tolerance limits. Many trace elements are not only harmful but toxic and fatal too. Among the earlier declared toxic elements a few like fluorine, bromine, barium, chromium and molybdenum have been lately found associated with different body functions. However, deficiency of these minerals is rarely recorded in practical feeding.

Excess of mineral in compounded feeds and drinking water requiring attention are aluminium, arsenic, barium, bromide, cadmium, chromium, copper, fluorine, lead, mercury, molybdenum, nickel, silver, strontium, sulphur, tunguston, selenium and vanadium. Some of the inorganic compounds of these elements are highly toxic and fatal. In recent years excessive use of under ground drilling, underground water and explosives is increasing the danger of toxic minerals pollution.

8. VITAMINS

Vitamins are highly unrelated chemical compounds present in the plants, animals and fungi, and essential for the maintenance of normal physiological functions.

The name was given by C. Funk in 1912 on observing cure of some diseases by eating specific foods. Original name was vitamine as Funk assumed that some amino-nitrogen is producing the beneficial effects which could not sustain later discoveries of many vitamins of diverse composition and name was changed to vitamin for representing the group of these organic micronutrients essential for normal physiological functions. Before the invention of recorded knowledge of vitamin it was known that certain factors in some plants and animals are capable of curing some diseases. Like any other animals, poultry are also affected by the imbalances particularly the deficiency of vitamins results in decreased production. Thus, commercial poultry farmer can not afford feeding of vitamins deficient diets.

There was great variation in the nomenclature of vitamins which was causing confusion in communication. This was rectified and resolved and published in1990.

Classification of vitamins

Vitamins are classified into two main groups of fat soluble and water soluble substances.

A. **Fat soluble vitamins:** Four fat soluble vitamins are A,D,E and K. Each vitamin is found in different forms and variable functions efficiency. Fat soluble vitamins are dietary essential but do not require daily supplementation because these can be stored in body fats for several days or much longer period.

B. **Water soluble vitamins:** These are not stored in the body except vitamin B12 and daily supply is required. Common feeds of poultry are satisfactory sources of many water soluble vitamins. The water soluble vitamins are thiamin (B1), riboflavin (B2), niacin or nicotinic acid, pantothenic acid, Vitamin B6 (Pyridoxine, pyridoxal and pyridoxamine), Vitamin B12 and folacin of the B-complex group and vitamin C or ascorbic acid.

A. FAT SOLUBLE VITAMINS

1. Vitamin A

Both preformed and precursors are available in nature and later forms are converted into vitamin A activity in the intestinal mucosa and liver. The rate of conversion of carotenoids into vitamin A activity is not similar in different species.

Vitamin A occurs in three forms.

(i) Alcohal form or retinol.

(ii) Aldehyde form or retinal

(iii) Acid form or retinoic acid or esterified form with a fatty acid.

All the three forms of active vitamin A are called retinoids and retinol is the most active form of Vitamin A.

Carotenoids or precurs of vitamin A are variable in vitamin A activity. Maximum conversion of carotenoids into vitamin A activity occurs in the intestinal mucosa followed by liver cells. Considerable proportions of carotenoids (mostly yellow to yellow orange in colour) remain unchanged and deposited in body fat. The carotenoids are carried in circulating fat and fatty acids. Egg yolk fat has high affinity for the carotenoids which impart yellow colour. Intensity of yellow colour of yolk depends on the source of concentration of carotenoids transfered from feeds into body of the laying birds. So far more than 500 types of carotenoids have been identified in plants. Till to day not a single plant has been found to possess active form of vitamin A. Largest proportion of absorbed carotenoids are stored in the liver and transported to ovary in blood circulation where greater amount is retained in the yolk of developing follicles during oogenesis.

Functions

1. One of the most important function of vitamin A is the formation of rhodopsin or visual puple essential for sight. Lack of rhodopsin formation in the absence of vitamin A causes night blindness (xerophthalmia) leading to total blindness on prolonged deficiency.
2. Required for the maintenance of integrity of the epithelial cells.
3. It is essential for normal skeletal growth by maintaining the functions of osteoblasts (bone forming cells) and osteoclasts (bone phagocytes).
4. It is required for normal cellular metabolism, protein synthesis and gene expression. Two proteins actively involved are the cytosolic retinoid binding protein (CRBP) and cytosolic retinoic acid binding protein (CRABP).
5. Maintenance of normal reproduction.
6. Regulation of immune system.
7. Protection from cancer.

Deficiency symptoms

1. Anorexia, ceasation of growth in chicks and loss of weight in adults.
2. Ailments of eyes-night blindness.
3. Rough and dull feathers.

4. Decreased immunity and increased risk of infection.
5. Metaplasia of nasal passage.
6. Lowered egg production and hatchability and high embryonic mortality.
7. High death due to secondary infections caused by damaged epithelium.

Sources of vitamin and its precursors

1. Fish oil like cod liver oil, shark liver oil etc.
2. Fish meal and fish trash.
3. Green and yellow leafy vegetables, carrot, pumpkin etc are rich in carotenoids.

Factors affecting functions

1. Protein and fat deficiency reduces uptake, transport and utilization of vitamin A (retinoids) and carotenoids.
2. Antioxidants, sugars, protein and lipids improve utilization.
3. Minerals particularly copper salts significantly destroy stability of vitamin.
4. Synthetic vitamin A supplements are considerably destroyed during pelleting and should be added 50 percent higher than the requirement.

2. Vitamin D

Synonyms of vitamin D are anti-rachitic vitamin, sun shine vitamin or sterol vitamins. Many sterols have been found to have vitamin D activity but only two forms are more useful. These are ergosterol or calciferol (D2) and 7-dehydrocholesterol (D3). Efficiency of utilization of vitamin D2 is very low in poultry and should be considered only when vitamin D3 is not available for supplementation. Vitmain D_2 is present in plants and D3 in animals.

Functions

1. It is essential for the intestinal absorption of calcium and phosphorus.
2. Maintenance of high Ca and P levels in plasma essential for normal skeletal growth.
3. Prevention of Ca deficiency tetany by mobilizing Ca to maintain plasma level. This occurs due to associated action of parathyroid hormone (PTH).
4. It is associated with the maintenance of immune system.

Deficiency symptoms

1. Rickets in chicks identified by sitting in squatting posture, reluctant to move, deformed bones, stiff gait, elastic bones, beak and deformed claws.
2. Drop in egg production.
3. Increased occurrence of thin shelled and shellless eggs.
4. Decreased hatchability because embryos are unable to break the shell due to weak beak and toes.

Sources

1. Exposure to sun light. One can observe foraging birds spreading their wings frequently for solar exposure. Some birds lie on back for some time.
2. Eating insects, snails and fish meal.
3. Supplementation of commercial vitamin D3 in feeds

3. Vitamin E

Vitamin E is an excellent anti-oxidant available from different common feeds of poultry. It is fat soluble oxygen scavenger and protects fats and fatty acids from oxidation. Chemically it is tocoferols found in different forms like alpha, beta and gamma etc. Alpha-tocoferol is most potent among the all forms. Some functions of vitamin E are also done by selenium but complete replacement with selenium is not possible.

Functions

1. As potent biological anti-oxidant inhibits production of free radicals from the oxidation of fatty acids and protects the damage of biological membranes.
2. It is associated with the synthesis of many specific proteins.
3. Synthesis of phospholipids necessary for the maintenance of cell membrane.
4. Provides protection to leukocytes and macrophages during infection and enhances disease resistance.
5. Maintenance and regularization of cell membrane permeability.
6. Increases antibody production for protection against diseases.
7. Harmful function is the increased toxic effects of heavy metals like cadmium and mercury.

Deficiency symptoms

1. Exudative diathesis. It is also prevented by Se.
2. Encephalomalacia (abnormal Purkinje cells in the cerebellum of brain) is caused by vitamin E deficiency and can not be prevented by Se. It is also known as crazy chick disease.
3. Reproductive failure manifested by embryonic degeneration during incubation. In male birds testicular atrophy occurs.
4. Nutritional myopathy or muscular dystrophy commonly observed in turkey poults and chicks.

Sources of vitamin E

1. Vegetable oils
2. Oil of cereal grains and their by-products.

4. Vitamin K

It is group of fat soluble compounds essential for prevention of bleeding by blood clotting. The two important compounds are phylloquinone (K1) and menaquinone (K2). Vitamin K1 is found in green plants and K2 is synthesized by enteric bacteria. All forms of absorbed vitamin K are converted into K2 or active form by liver.

Functions

Synthesis of prothrombin or blood clotting factor II in liver, which is essential for blood clotting.

Deficiency symptoms

Main cause of vitamin K deficiency in poultry is the occurrence of coccidiosis and its treatment with sulpha-drugs. Mycotoxins, dicumarol, arsenilic acid and warferin like products are also responsible for vitamin K deficiency. Young chicks are highly susceptible and death due to bleeding is frequently observed. Some of the common symptoms are:

(i) Erosion of gizzard.

(ii) Mortality of chicks due to internal haemorrhage.

(iii) High chick mortality in hatches of vitamin K deficient hens.

Common sources

1. Cereal grains and their oil containing by products.
2. Oil containing cakes also contain small amount of vitamin K.
3. Liver meal and fish meal.
4. Coprophagy in foraging birds.

B. WATER SOLUBLE VITAMINS

The vitamins of B complex group and ascorbic acid (Vitamin C) are water soluble and dietary essential because these are neither synthesized nor stored in the body.

1. Thiamin (Vitamin B_1)

It is a compound of one molecule each of pyrimidine and thiazole. Earlier names are aneurin, vitamin F and thiamine. It has main role in the formation of co-enzymes necessary for energy metabolism.

Functions

1. Release of energy from nutrients during oxidation in the body tissues. Thiamin is converted to thiamin pyrophosphate (TPP) or co-carboxylase in the liver by phosphorylation. This coenzyme is involved in decarboxylation of keto-acids.
2. Synthesis of nucleic acid through pentose phosphate cycle.
3. Maintenance of membrane integrity and function of nerve cells.
4. Synthesis of acetyl choline and fatty acids.
5. Release of energy for body functions.
6. Transportation of sodium ions to nerves.

Deficiency symptoms

1. Development of polyneuritis in poultry. It is caused by incomplete metabolism in thiamin deficiency resulting in accumulation of incomplete metabolites of carbodhydrates around the nerves which damage nerve cells.
2. Loss of appetite and digestive disorders result in body weight loss, lowered production and general weakness.
3. Ophisthotonus or star-gazing due to nervous degeneration.

4. Reproductive atrophy in chronic deficiency.
5. Enlargement of heart, oedema of dependent parts and decreased heart beat.

2. Riboflavin (Vitamin B_2)

It is a yellow florescent pigment. It is a complex compound of ribose sugar and iso-alloxazine. Earlier names, not in use, are lactoflavine, riboflavine or vitamin G. The vitamin is heat stable but destroyed on exposure to light. Riboflavin activity is present in natural products like flavin mononucleotide (FMN) and flavin di-nucleotide (FND).

Functions

1. It is a functional unit of coenzymes FAN and FAD required in many enzyme systems of metabolism pertaining to transfer of electrons in oxidation-reduction reactions.
2. It is essential for release of energy from the metabolism of carbohydrates, fats and proteins.
3. It is associated with oxidation reaction in cornea.

Deficiency symptoms

1. Curled toe paralysis in chicks due to peripheral nerve degeneration. Inwards curling of toes causes difficulty in movement and chicks walk on hocks.
2. Retarded growth but appetite is normal.'
3. Reduced hatchability.
4. Abnormal embryonic development like "Clubbed down" formation characterized by growth of coiled feathering caused by growth of feather inside the follicles. Formation of dwarf embryo.
5. Diarrhoea and death.
6. Drop in egg production and hatchability.

3. Niacin or Nicotinic acid

Several pyridine compounds possess niacin activities and it has been agreed that the term niacin will represent nicotinamides. Earlier name is vitamin PP.

Functions

1. Niacin is a component of coenzymes, nicotinamide adenine dinucleotide (NAD) and nicotinamide adenine dinuleotide phosphate (NADP). These are essential for many important functions in the body like transfer of hydrogen from nutrients to molecular oxygen for water formation.
2. These two coenzymes are involved in many enzymic reactions for the metabolism of carbohydrates, fat and proteins for the release of energy.
3. Associated with synthesis of nucleic acid via hexose monophosphate shunt in the form of NADP.

Deficiency symptoms

1. Bowing of legs due to enlargement of tibio-tarsal joints.
2. Poor feather development
3. Dermatitis of feet and head.
4. Black tongue due to inflammation of oral cavity involving tongue, inner surface of mouth, pharynx and adjoining oesophagus.
5. Anorexia and poor performance of chicks.
6. Anorexia, decreased egg production and poor hatchability in layers.

Common sources

1. Niacin in feeds of plant origin is present in bound form and sparingly available
2. Animal Protein supplements are good sources
3. Some requirement of niacin may be met by conversion of tryptophane, if provided in excess of normal requirement for protein synthesis.

4. Pantothenic acid

Pantothenic acid is available in the forms of calcium and sodium salts. Former is preferred for mixing in feeds due to less hygroscopic property than the sodium salt.

Functions

1. It functions in metabolism as co-enzyme A and acetyl carrier protein.
2. Improves antibody concentration in blood by incorporation of amino acids in albumin.

3. Co-enzyme-A is essential for the release of energy from the oxidation of fatty acids.
4. Essential for synthesis of acetyl choline, the chemical transmitter at the nerve synapse.
5. Involved in synthesis of haemoglobin and citric acid.

Deficiency symptoms

1. Loss of appetite
2. Poor growth and low feed conversion efficiency.
3. Poor feather formation.
4. Dermatitis at the beak commissures. Cracks and fissures may be observed on toes and legs.
5. Wart like lesions may be seen on legs. This is aggravated by biotin deficiency.
6. During the terminal period of incubation high embryonic loss occurs due to oedema and subcutaneous haemorrhages.

5. Biotin (Vitamin H)

Initially it was found as a growth factor for yeast and later on its role in animal body functions was investigated.

Functions

Biotin is required for the functions of many enzyme systems.

1. Conversion of carbohydrates and proteins to fat.
2. Conversion of carbohydrates to proteins and vice-versa as per requirement in the body.
3. It is essential for the maintenance of blood glucose level.
4. Biotin is involved in haemoglobin synthesis.
5. It is considered to be associated with aspartic acid synthesis.

Deficiency symptoms

The functions of biotin are adversely affected by the presence of moulds in the feeds and occurrence of rancidity in badly stored feeds.

1. Reduced feed intake, poor growth and lowered feed conversion efficiency in growing chicks.
2. Development of deformities like ataxia, perosis, crooked legs and parrot beak.
3. Broken and deformed feather.
4. Dermatitis produced in pantothenic acid deficiency is aggravated by biotin deficiency.
5. Hatchability is lowered in breeding flock.

6. Vitamin B6

Since three compounds, viz. pyridoxine, pyridoxamine and pyridoxal are involved in the same body functions, these should be called vitamin B6.

Functions

Vitamin B_6 works as a component of Co-enzyme in many enzyme systems in the body.

1. It is associated with synthesis as well as catabolism of the amino acids.
2. Essential for the release of energy from the metabolism of carbohydrates, lipids and proteins.
3. Required in reactions of tricarboxylic acid cycle.

Deficiency symptoms

1. Loss of appetite, reduced growth and feed conversion efficiency in growing chicks.
2. Hyperexcitability and restlessness in severe cases. Birds run aimlessly with characteristic posture of slightly downwards wings.
3. Violent convulsions and death in severe and prolonged deficiency.
4. Extensive erosion in gizzard.
5. Reduced egg production and hatchability.

7. Folacin

Since the vitamin activities are attributed by folic acid and other compounds, earlier name folic acid was derived from foliase because it is found in green leaves. Some of the derivatives of folic acid are tetrahydrofolic acid, 5-methyl-tetrohydrofolic acid, 5-formyltetrahydrofolic acid and 10-formyltetrahydrofolic acid.

Functions

1. Transfer of single carbon unit and its incorporation in larger molecules.
2. Tetrahydrofolic acid is used for the production of purine, pyrimidine, glycine, serine and creatine.
3. Synthesis of oxidases like choline oxidase and xanthine oxidase associated with choline and methionine metabolism.
4. Normal feather pigmentation in coloured breeds.
5. Maintenance of immune system of the body.

Deficiency symptoms

1. Macrocytic hyperchromic anaemia, leucopenia and throbocytopenia.
2. Reduced growth and poor feather growth in growing chicks.
3. Depigementation of feather and skin in coloured breeds.
4. Abnormal formation of hyaline cartilage and poor ossification of bones resulting in perosis.
5. Decreased hatchability and increased embryonic loss.

8. Vitamin B_{12} (Cyanocobalamin)

Old name of this vitamin is animal protein factor (APF) due to occurrence of anaemia in animals on the feeding of animal protein free diets. This vitamin is synthesized by bacteria, if cobalt is available. Folacin has also role in vitamin B_{12} synthesis. Stability is good and activity is maintained in minerals containing compounded feeds and pellets.

Functions

1. It is a co-enzyme in many enzyme systems and participate in metabolism.
2. Involved in synthesis of methionine; and folacin is required in the reaction.
3. Inter conversion of amino acids.
4. Involved in synthesis of purine, pyrimidine and proteins.

Deficiency symptoms

1. Loss of appetite, reduced growth rate, poor feathering and lowered feed conversion efficiency.

2. Kidney damage, thyroid dysfunction and perosis.
3. Depression of plasma proteins level.
4. Elevation of blood glucose and nonprotein nitrogenous substances.
5. Lowered hatchability. Embryonic death around day 17 of incubation. Dead embryos show multiple haemorrhages, enlargement of heart and thyroid glands, fatty liver and absence of myelin sheath on the nerves specially the sciatic nerve and the spinal cord.

9. Choline

It is found in animal tissues as choline and acetylcholine. Choline is highly soluble.

Functions

1. It is structural constituent of lecithin and sphingomyelin.
2. Associated with transmission of nerve impulse.
3. Prevention of hepatic fattiness by mobilizing fat as lecithin and increasing fatty acids metabolism.
4. Essential component of acetyl choline.
5. Labile methyl group donar action involved in methionine synthesis from homocystine.
6. Required for the normal maturation of the cartilage matrix of bones in the form of phospholipids.
7. Involved in peripheral vasodilation, contraction of skeletal muscles and regulation of heart beets.

Deficiency symptoms

1. Perosis in chicks similar to that observed in manganese and biotin deficiency.
2. Development of fatty liver. This may be prevented by excess supplementation of methionine.
3. May affect egg size in layers.

10. Ascorbic acid (Vitamin C)

The unknown factor in lemon and orange that cured scurvy centuries back was named vitamin C. The ‘C’ was derived from citrus. Later on it was found to be ascorbic acid. This is water soluble vitamin. It is found in two chemical forms, reduced form or ascorbic acid and oxidized form or dehydroascorbic acid.

Ascorbic acid is synthesized in the body of poultry and many other birds except the red vent bulbul and yellow vent bulbul. There is remote chances of vitamin C deficiency even in the bulbuls due to opportunity of feeding on citrus fruits.

Functions

1. It is component of enzymes involved as a catalyst in many oxidation-reduction reactions in the body.
2. It is a reducing agent.
3. Associated with electron transfer in cells.
4. It is essential for conversion of tetrahydrofolic acid from folacin and utilization of vitamin B12. These are necessary for the maintenance of haemoglobin and prevention of anaemia.
5. Stimulates phagocytosis and antibodies production.
6. Alongwith ATP it is involved in incorporation of plasma iron into ferritin.
7. It is associated with normal bone development.
8. Essential for steroids synthesis.
9. Facilitates absorption, transportation and distribution of metal ions in the body.
10. Required for the synthesis of carnitine used for the fat metabolism.

Deficiency symptoms

Adequate amount of vitamin C is synthesized even in the day-old chicks and normally dietary supplementation is not required. However, stress conditions may create temporary deficiency requiring short duration supplementation.

5

FEED ADDITIVES

Earlier feed additives were limited to non-nutrient constituents added in minute amount in the feed or drinking water of poultry for stimulating growth and egg production and also providing protection against certain ailments. The earlier feed additives were antibiotics, arsenicals, hormones, phytohormones and surfectants.

The area of feed additives has broadened and now feed additives are non nutrients or nutrients, organic or inorganic, natural or synthetic or complex non toxic compounds used for enhancing growth, production, quality of products and maintenance of good health.

The use of such substances will normally not have residual effects on consumers and the additives, their residues and metabolites in the foods of animal origin will not have any adverse effects on the consumers. It may be also included that the spilled additives and their metabolites voided in excreta and exhaled air will not have environment polluting effect.

At present the family of feed additives has grown complex and continuing to grow every day. Now it has become an aggregation of substances of diversified origin and chemical composition capable of stimulating growth and egg production, improving feed conversion efficiency, and maintaining good health. These will also not have any kind of adverse or harmful effects on the consumers of products like poultry meat and eggs.

Different kinds of feed additives

Different kinds of feed additives with beneficial claims in poultry production are available in the market. These feed additives including neutraceuticals may be classified into the following groups.

1. Antibiotics
2. Anti-coccidials

3. Anti-oxidants
4. Anti-caking substances
5. Aromatic compounds or flavour producing substances
6. Arsenicals
7. Adsorbants
8. Colouring agents
9. Emulsifier
10. Enzymes
11. Grits for gizzard health
12. Herbal products
13. Hormones and phytohormones
14. Mould inhibitors
15. Prebiotics like mannan, oligosaccharides
16. Probiotics
17. Synbiotics
18. Proteinates and mineral chelates
19. Synthetic amino acids
20. Essential fatty acids

Points to be considered for the selection of feed additive (s) for use in poultry production.

The primary purpose of poultry farming is the production of table eggs and edible flesh. Both these products should be wholesome and fit for human consumption. Some of the feed additives may remain residue in flesh or passed in the edible contents of eggs. Therefore, it should be considered essential point to keep in mind the effects of various additives on human health before the incorporation of such additives in the diets of poultry.

1. The products should be safe for poultry birds and human consumers of poultry products.
2. The residue should not accumaulate in the body tissues or passed into the eggs.
3. Should not void harmful residue and metabolites.
4. Should not be immuno depresent
5. Should not be carcinogenic

6. Should be economical and cost effective.
7. Should not create environmental pollution problems.

1. Antibiotics

Some antibiotics are still used for enhancing growth and feed utilization efficiency of growing birds and fattening birds. Antibiotics are normally beneficial in birds reared in poor hygienic conditions particularly the microbial invasion of the alimentary canal. Pathogens in gut destroy the mucosal integrity and reduces digestibility and absorption of nutrients. Antibiotics work as bactericidal or bacteriostatic and improve the digestion and absorption of the nutrients. These have glucose and amino acids sparing effects by reducing or preventing the lactic acid production and harmful amines respectively. Sub clinical presence of pathogens in digestive tract produce harmful substances like cadaverine and putrescine in the caecal contents.

Feeding of antibiotics should be discontinued during the finishing period of broilers and should not be used in the diets of layers and the birds finshied in short duration of rearing like quail, pigeon and partridge for meat production.

Common antibiotic feed additives for poultry: Avilamycin, flavomycin, lincomycin, zinc bacitracin, Virginiamycin, Oxytetracylcline and chlortetracycline.

Use of antibiotics as feed additives has been banned in many countries and should be banned in the remaining countries. The reasons are the (i) development of antibiotic resistant pathogens challenging human health and evolution of antibiotic resistant pathogenic strains of many disease. Inhibitory legislation may be imposed in countries like India where almost half of the population is unable to bear the burden of hospitalization.

2. Anticoccidials

Cocidiosis caused by Eimeria spp. is a common problem in poultry production and the infection is more in bad hygienic conditions, deep litter rearing and backyard production. The anti-coccidial drugs should be used for the birds reared in sub optimal conditions of housing and also in the diets of foraging birds kept in semi-extensive and extensive systems.

Coccidia damage gut mucosa resulting in decreased absorption of nutrients from the digestive tract. Now anticoccidial vaccine is available and used at large broiler breeder farms for the vaccination of parent flock. For the birds reared in deep litter housing anticoccidial drugs are used for prevention.

Some of the antii-coccidial drugs are clopidol, 3, 5,- dinitro toluamide (DOT), robenidine, amprolium, arpinocid, nicarbazine, quinoline, metichlorpindol, halofuginone, ionophores like monensine, madhuramycin, salinomycine and lasolacid, sulphamezathine, few herbal products, sapoginin and new product diclauzuril etc. Development of resistance against anti-coccidial drugs is a serious problem in prevention of coccidiosis and requires necessary change in the use of drugs of different composition for the prevention of coccidiosis.

Change of litter material, frequent turning of litter and housing in cages in battery system are some of the effective measures for the control of infection. Chicks should also be produced from the coccidia free parent flocks.

3. Anti-oxidants

Oil rich feeds like rice polish, rice brans and other feeds are frequently used for the compounding of poultry feeds. Almost all such oil containing feeds are prone to auto-oxidation resulting in development of rancidity spoiling flavour, taste and nutrients availability. High atmospheric temperature and humidity favour feed breakdown and liberation of highly reactive free radicals. The rate of oxidative rancicity is higher in crushed feeds and compounded feeds containing mineral supplements. Iron and copper are pro-oxidants and catalyse oxidative rancidity in the compounded feed mixtures.

The antinutritional or harmful effects of rancidity may be listed as follows:

(a) Voluntary feed intake is reduced due to decreased palatability.

(b) Availability of nutrients from the feed is reduced.

(c) Reduction in pigmentation of skin and egg yolk due to oxidation of dietary carotenoids.

(d) Destruction of activities of fat soluble vitamins (A, D, E) and also biotin.

(e) Spoilage of flavour of feeds and poultry products.

(f) Economic loss due to commulative effects.

Both natural and synthetic antioxidants are available which prevent oxidative rancidity. Anti-oxidants are also immuno-stimulant. Use of synthetic anti-oxidants spare natural vitamin E in feeds for performing the physiological functions.

Some of the synthetic anti-oxidants and their quantity mixed in feeds for effective anti-oxidant effects are presented in Table. 5.1.

Table 5.1: Quantity of synthetic anti-iosxidants for preventing oxidative rancidity.

Anti-oxidants	Amount per quantial
Butylated hydroxyl anisole (BHA)	12.5 g or more
Butylated hydroxyl toluene (BHT)	12.5 g or more
Ethoxyquin	12.5 g or more
Mixed commercial anti-oxidantsAntioxidatns –cum-nutriets	
Selenium	10 – 15 g
Vitamin C (Ascorbic acid)	10 – 100 g
Vitamin E as alpha tocopherol	1.5 – 3 g

4. Anti-caking substances

Substances used in small quantity for checking the lump formation in feeds are the anti-caking substances. Attapulgite (maximum 0.25% of dry matter) and bentonite (maximum 2.5% of dry matter) are satisfactory anti-caking factors.

5. Aromatic compounds or flavoring agents

Certain aromatic substances are used to increase the flavour and palatability of feeds. These are available in powder and liquid forms. Liquid aromatic agents are misible in water or lipids. These substances are preferably used to improve the flavour of feeds but also used to mask the unpalatable aroma in the feeds.

6. Arsenicals

Some compounds of arsenic are used in very small quantity for improving the performance of birds. Earlier arsonetic acid was used. At present 3-nitro-4-hydroxyphenyl arsonic acid (below 0.01%) is used. Arsenic accumulates in liver and should be withdrawn from the diet of broilers and other meat birds one week before sacrifice or marketing.

7. Adsorbents

Some substances are capable of adsorbing toxins and harmful substances, preventing their absorption in the body tissues and eliminating in the faeces are known as adsorbents. The reaction is a physical affinity or physical binding. In recent time use of adsorbents is increasing due to increase of mycotoxins and residual pesticides in feed ingredients. Some of the common adsorbents are activated charcoal, alumino silicates, bentonites and zeolites. A common complex compound used as adsorbent is hydrated sodium calcium aluminium silicate(HSCAS). It is a very effective binder of aflatoxins and can bound upto 100 ppb from the feeds. Although aflatoxins binding capacity of HSCAS is about 500 ppb in vitro which is drastically reduced by several factors in actual feeding.

Some factors adversely affecting the adsorption efficiency of HSCAS are the required time for adsorption, rate of release of bound aflatoxins during peristalsis, gradual release of deeply dispersed aflatoxins in the feed ingredients.

HSCAS does not bind other mycotixins like citrinin, ochratoxin and sterigmatocystin and residues of pesticides and fungicides in the feeds.

Activated charcoal is a broad spectrum adsorbent capable of binding most of the toxins in feeds and alimentary canal like mycotoxins, pesticides and fungicide residues.

A mixture of activated charcoal and HSCAS can be a more effective adsorbent due to synergetic action.

These adsorbents had also some negative effects due to binding of some nutrients. Activated charcoal binds vitamins due to which dietary requirements is increased. Similarly HSCAS binds calcium and increases dietary requirement. Mannon oligosaccharide (MOS) has been found to bind toxins and also eliminated pathogenic organisms from the digestive tract.

8. Colouring agents or pigments

Orange yellow or reddish yellow yolk colour has wide acceptance and significant dilution or change of yok colour to faint yellow or dull white retracts customer. Yolk colour is diluted drastically on replacement of yellow maize with wheat, rice, white sorghum, cassava flour and minor millets. Now some natural source of yellow colour like turmeric powder, annato power and leaf meal are being used in compounded feeds for imparting yellow colour to yolk. The practice is acceptable till recommended colours fit for use in human foods are used. But, use of metallic colour and extracts of harmful herbages like powder of yellow Kaner flower will be detrimental for health and must not be used.

9. Emulsifier

The substances used for the emulisification of fat are known as emulsifiers. Emulsification increases surface area of fat globules and dispersal which increase digestion and absorption. The development of fast growing broiler strains has significantly increased the requirements of energy and other nutrients which are to be supplied in the limited capacity of alimentary canal. For increasing the metabolizable energy concentration in compounded feeds about 2 to 5 per cent fats are mixed. Chicks are also not fully able to utilize dietary fat in first week of like. Use of emusifiers enhances dietary fat digestion and absorption. Lecithins are good emulsifiers and lecithin of soyabean is commonly used in poultry feeds.

10. Enzymes

Cereal by products like deoiled rice bran and wheat bran and oilseed cakes/ meals consititute about half of the composite diets of poultry. These contain considerable amount of non starch polysaccharides (NSPs) which are cellulose, hemicellulose, pectin, galactosides and beta-glucans etc. The cellulose is a polymer of glucose and hemicelluloses also contain good amount of glucose alongwith other sugars. The energy present in cellulose and hemicelluloses are not available to poultry due to non availability of necessary enzymes required for digestion in the pre-caecal parts of the digestive system.

Cell wall of plant tissues contains NSPs which are disintegrated by crushing and soaking in gizzard. The nutrients are released and area increases for digestion in the lower tract. Some of the NSPs like pectin and beta glucans absorb large quantity of water resulting in swelling and increasing the viscosity of the gut contents. This causes hinderance in enzyme action and absorption of the nutrients.

The mineral constituents of the plant feeds are present mostly as phytate (a hexaphosphoric acid ester of inositol) which is not available to poultry due to lack of phytase, an enzyme essential for the release of phosphorus from phytate form. Phytates in feeds are also anti-nutritional factor and decreases the absorption of other nutrients like proteins, amino acids and several minerals like calcium, magnesium, copper, iron, manganese and zinc due to formation of phytates.

The manner of action of additive enzymes on the release and availability of nutrients has been observed to work as (i) the phytase is used for the disintegration of phytates and liberaction of proteins, amino acids, phosphorus, calcium and other minerals (ii) Cellulases are used for the rupture of cell wall and liberation of carbohydrates, proteins and lipids to be digested by the digestive enzymes secreted by poultry. This results in upto 10 percent higher gain and feed conversion efficiency of poultry (iii) The constituents of cell walls of cereal grains like beta-glucans and arabinoxylans are not degraded by endogenous digestive enzymes of poultry and increase viscosity resulting in less availability of nutrients and pasty nature of the excreta. Sticky excreta soils feathers of the birds and cages, and create problem in excreta collection and disposal. Supplementation of glucanase enzyme in barley containing diets significantly improves digestibility and availability of energy and other nutrients. The response of growing chicks has been reported comparable with maize based diets. Thus, exogenous enzyme additives have been found to increase the availability of bound nutrients resulting in higher growth, improved nutrients conversion and ultimately higher return form the products.

Various enzyme feed additives of microbial sources capable of breaking down NSP molecules are now available commercially and have been found to increase availability of nutrients, performance of birds and less problem in manure management. These are phytase and non starch polysaccharidases like cellulases, pectinases, beta-glucanases, hemicellulases and arabinose xylanases.

The actions of enzyme feed additives may be summarized as follows:

1. Rupturing the cells for increasing enzyme action and liberation of nutrients for digestion.
2. Increasing rate of saccharification of NSPs.
3. Boosting the action of endogenous enzymes.
4. Enhancing production performances (growth and egg production) by increasing nutrients availability.
5. Decreasing quantity and improving consistency of excreta for convenient disposal.
6. Overall improvement in the health of poultry resulting in increased economic returns.

The enzymes should function in wide range of pH (2 to 7), should have reasonably short reaction time to cope with the feed retainsion time, can be stored at room temperature, can sustain temperature rise during feed processing, should be economical and easily available.

11. Grits for gizzard health

Stone or marble grits are often offered free choice although not essential. Earlier it was considered necessary because hard pieces (grits, metals etc.) were frequently found in the gizzard of scavenging and foraging birds. Use of shell grits and marble grits may be beneficial and laying hens will be able to meet higher calcium demand for egg shell synthesis by free choice consumption of shell grits or marble grits.

12. Herbal products

Many herbs and herbal products have been found to increase palatability causing increased feed intake more nutrients availability and resultant higher production. The other merits are antioxidant and immunogenic nature. Use of herbs and herbal preparations for the protection of health and enhancement of productivity is considered safe, economical, easily available and eco-friendly. Advantages of herbal feed additives may be listed as follows:

(a) Increases feed intake and nutrients utilization efficiency.

(b) Quantitative as well as qualitative improvement in production.

(c) Maintenance of good health and protection to liver from harmful substances

(d) Protection of renal system against nephrotoxins.

(e) Alleviation of stress.

(f) Immunomodulation

(g) Antioxidant effect due to presence of phenolics.

(h) Phenolic compounds are antibacterial and provide protection against respiratory infections. Phenols increase cell permeability resulting in increased water imbalance and bacterial cell death. This phenomenon does not allow development of resistance against phenolic compounds.

(i) Some herbal products have coccidiostatic effect: Some of the herbal products used as such or after extraction of active principles are aloe, cinnamon, cardamom, basil, garlic, ginger, mint, thyme, sage, cumin and clove etc. Mixtures of some of such ingredients are also available. However, more information generation is necessary for identification of specific benefits.

13. Hormones and phytohormones

Oestrogenic and somatotropic hormones were used for increasing growth and fat deposition in the carcass during the previous century. Later on it was found to leave residue in the meat and egg that was claimed to be deliterious for human health. Now use of hormones as feed additives in the diets of food animals has been banned in most of the countries.

14. Mould inhibitors

These are substances used in small quantity for the prevention of mould growth and spoilage of stored feed ingredients and compounded feeds. Moisture favours mould growth which is enhanced by high temperature. In order to prevent mould growth the feeds are sprayed with dilute solution of one of the organic acids capable of inhibiting implantation and growth of mould spores. Some of the common mould inhibitors are acetic acid, benzoic acid, formic acid, propionic acid and sorbic acid. The fungicidal action of these organic acids is more effective for the protection of cereal grains, their by products and starchy foods. The effect is less effective on proteinous feeds like oil cakes and fish meal and alkaline minerals.

Mould inhibitors should be changed at proper interval to avoid the development of resistance in fungi or mutation of fungi due to prolonged use. Dose of mould inhibitors should not be sublethal to control growth of fungi and production of fungal toxins. Effect of copper sulphate is less on mould growth but effective on inhibiting the synthesis of aflatoxins. For uniform distribution and inhibition effect on mould growth finely powdered copper sulphate is mixed with finely powdered common salt in 1:10 ratio. Gention violet is an effective mould inhibitor but it is carcinogenic and banned.

15. Prebiotic or Mannan oligosachharides (MOS)

Certain non starch polysaccharides (NSPs) present in common poultry feeds have been claimed beneficial for poultry production. These NSPs are indigestible in chicken. These NSPs reach caeca and lower tract unaltered and increase multiplication rate of one or more useful bacteria in the large intestines. This helps improvement of gut health as well as host health. These non-starch polysaccharides are also called prebiotics and mannan oligopolysaccharides (MOS) are more effective. Others are galacto oligosaccharides, fructo-oligosaccharides and lactose derivatives. Sources of different oligosaccharides are given in table 5.2. Some useful oligosaccharides are also manufactured and marketed for use as prebiotics in the diets of simple stomached animals.

Table 5.2: Sources of different oligosaccharides

Oligosaccharides	Common natural sources (foods)
α-galacto-oligosaccharides (GOS)	Leguminous seeds, rapeseed mustard, soyabean, other beans
Fructo-oligosaccharides (FOS)	Cereal grains and their brans
Mannan oligosaccharides (MOS)	Yeast cells and commercial
Trans galacto-oligosaccharides (TOS)	Milk products

16. Probiotics

Probiotics are beneficial live microorganisms that modify gut microflora for favouring multiplication of useful strains or species of gut microbiota by suppressing the growth of low priority microorganisms. The alimentary canal of avian species under normal conditions is sterile at hatching but acquire certain species of microorganisms from other birds and surroundings. These microorganisms become integral member and inhabit generally in the lower gut. These microflora play complementary role by synthesizing digestive enzymes under normal conditions. Now it has been observed that introduction and establishment of selected beneficial live microorganisms encourage the prolification of beneficial strains of the gut and had additive effects on the availability of nutrients and productivity of the poultry. Increased nutrients availability is the effects of higher digestion and absorption of the digested nutrients.

Characteristics of direct fed microorganisms: For effective performance the direct fed microorganism should posses the following properties:

1. It should be capable to survive and multiply amidst the already inhabiting gut microflora.
2. It should be capable to balance the desirable gut microflora.
3. Probiotic (Direct fed live microflora) should not be harmful for the host.
4. These should be resistant to acid (In proventriculus) and bile draining into duodenum.
5. Probiotics should be able to destroy or control the prolification of harmful pathogenic microbes.
6. These should be resistant to stress of feed processing and feed storage conditions.

Probiotics or Direct fed live microbials: Used in poultry feeding are both bacteria and fungi. Some of the bacteria are Lactobacillus acidophilus, L. bifidus, L. brevis, L. bulgaricus, L. cellobiosus, L. cremoris, L. fermentum, L. plantarum, L. reutero, L. salivarius, L. sporogenes, Pdiococcus halophilus, P. pentosaccus, Streptococcus diacetylactis, Str. faecium, Str. lactis and Str. thermophilus. Some others are Bacillus cereus and B. subtilis. The probiotic fungi are the strains of Aspergillus oryzae, Saccharomyces cereviciae and S. boulardii.

Mode of action of probitics: Various ways of action of probiotics may be described as follows:

1. Bacteriostatic and bactericidal action helps to control pathogens through elimination in excreta. This is done by Lactobacilli. Few strains of Lactobacillus and streptococcus also produce antibiotic like substances that have bactericidal prioperty and kill pathogens.

2. Harmful toxins like aflatoxins are ingested by Lactobacillus species.
3. Adherence and proliferation of direct fed microbials on gut wall prohibit the colonization of harmful bacteria like Escherichia coli (E. coli). Bacteria posses many microfimbriae of proteinous substances, lectins. The lectins help in identification for attachment with specific oligosaccharide receptor sites on the gut wall. Lactobacilli are most effective for the removal of pathogens.
4. Inhibition of harmful and irritant amine production from amino acids by E. coli because the microbe is killed by the metabolites (lactic acid and antibiotic like substances) of Lactobacilli and Streptococci.
5. Increases immunity in birds.

Factors affecting the usefulness of probiotics: The following factors may affect the usefulness of probiotics:

(i) Age of the bird. More beneficial effects are found in growing chicks.

(ii) Housing and environment: Beneficial effects are more appreciable in unhygienic management.

(iii) Composition of diet and method of feed processing.

(iv) Number of viable cells of probiotic administered.

(v) Composition of probitics, i.e., single or mixture and mixture of bacteria or mixture of bacteria and fungi.

17. Synbiotics

These are synergistic combinations of prebiotics and probiotics used for enhancing the production performance and maintenance of good health. Synbiotics are still in the process of development and in future definition may be further elaborated.

18. Proteinates or mineral chelates

It has been often claimed that bioavailabilty of chelated or protein bound minerals is higher than the inorganic salts. The results are not consistant and economically viable for most of such proteinates. However, proteinates also contain amino acids and trace minerals proteinates with limiting essential amino acids may serve the dual purpose of balancing amino acids and supplying trace mineral.

19. Synthetic amino acids

Normally compounded poultry diets of cereal grains and oilseed cakes are found deficient in some of the essential amino acids like lysine, methionine, phenyalanine and threonine. Generally fish meal and meat meal are used in small quantity for balanced supply of essential amino acids. Now synthetic amino acids are available at a competitive price and used successfully for the compounding of balanced poultry feeds.

20. Essential fatty acids (Polyunasaturated fatty acids (PUFA) or omega fatty acids (Omega 3 and Omega 6)

Some polyunsaturated fatty acids (PUFA) are not synthesized by the chicken and require supplementation for optimum production. Linoleic acid, gama-linolenic acid and arachidonic acid are the essential fatty acids but only linoleic acid is essential for poultry. Deficiency of linoleic acid results in unthriftness, retarded growth, lowered egg production, smaller egg size and lowered fertility and hatchability.

Care should be taken in selection and supplementation of feed additives. During use of combinations of two or more feed additives careful selection is important and the components selected for the mixture should not be antagonistic. Synergistic effects are desirable.

6 ANTI-NUTRITIONAL FACTORS IN POULTRY FEED INGREDIENTS

The natural constituent(s) in a feed ingredient producing adverse effects on feed intake, digestibility, absorption and assimilation are anti-nutritional factors. Most of the nitrogenous and saccharid feeds of plant contain one or more anti-nutritional constituents.

Characteristics of anti-nutritional factors in feed ingredients

Different kinds of anti-nutritional factors are found even in the common feed ingredients of poultry and other farm animals. The chemical composition and properties of these substances are highly variable and may be acidic, alkaline or neutral in nature.

1. These substances had adverse effects on voluntary intake, digestibility, absorption and metabolism.
2. Some factors produce anti nutritional effects and in some feeds their metabolites produce harmful or even toxic effects.
3. Many anti-nutritional factors are unstable and either destroyed or inactivated by simple processing techniques like drying, boiling, roasting, pressure cooking, soaking and fermentation etc.
4. Few factors turn toxic during digestion due to change in chemical composition viz. Feed nitrates are reduced to nitrites which are toxic and may be fatal on crossing threshold. The nitrites bind with the hemoglobin of erythrocytes and converts ferrous form of iron in haemoglobin to ferric form in brownish methaemoglobin. Methaenoglobin is not capable of transporting oxygen to different tissues causing anoxia and death. Scope of nitrate/ nitrite poisoning in poultry is difficult from feed sources.

Different kinds of anti-nutritional factors in feeds

Different kinds of anti-nutritional constituents of the various feedstuffs and also the adulterants of feeds like mixing of argemone seeds with rape sees-mustard have been identified and listed in Table. 6.1

Table 6.1: Anti-nutrient factors in conventional and non-conventional feeds.

S.No.	Anti-nutrient factors	Feed sources
1.	Protease inhibitors (eg. Trypsin inhibitor)	Leguminous seeds- soybeans, cluster bean seeds, field bean, peas, lentil
2.	Phenols (i) Gossypol	Cotton seed and cotton seed cake.
	(ii) Tannins	mustard-rape seed cake sal seed meal, mango seed, babool seed, black bery (jamum) seed, sorghum, etc.
3.	Phytates	All feeds of plant origin.
4.	Oxalates	Sesame seed meal, many plant and animal feeds.
5.	Nimbidins	Neem seed and neem seed meal.
6.	Nitrates and nitrites	Immature plants, chemical contamination.
7.	Non-starch polysaccharides (NSPs)	Cereal grains, brans and vegetable proteins.
8.	Mimosine	Leucaena (koobabul / subabul) leaf meal.
9.	Glycosides	
a.	Cynogenic	Linseed meal, tapioca root.
b.	Glucocynolates	Mustard-rape seed meal
c.	Oestrogens	Soyabean meal, kidney been.
d.	Saponins	Soyabean meal, lucorne leaf meal, mahua seed meal.
10.	Haemagglutinins or lectins	Castor beans, cluster bean meal
11.	Erucic acid	Mustard-rape seed.
12.	Argemone	Adultrant in mustard-rape seed meal.
13.	Antivitamin factors	
a.	Antivitamin A	Soyabean seed (Lipoxygenase enzyme)
b.	Antivitamin D	Soyabeans
c.	Antivitamin E	Kidney bean
d.	Antivitamin K	Dicumarol (false vitamins K synthesis) in sweet clover
e.	Anti-vitamin B1 or Antithiamyne	Linseed meal
f.	Antipyridoxine	Linseed (linamarin)

Harmful effects of various anti-nutritional factors

The informations about the various anti-nutritional factors naturally occurring as constituents of many common feed resources and their deleterious effects on the health and production of poultry are important. These informations are used for the selection of suitable feed stuffs and also application of suitable treatmens for the inactivation / detoxification of such feeds. Deleterious effects of different anti-nutritional factors on the health and production of poultry birds need attention during selection of feeds and formulation of compounded diets.

1. **Protease inhibitors (Trypsin inhibitors):** These are harmful constituents present in many vegetable feeds. So far about six protease inhibitors have been found. The important ones are the Bownem–Birk chymotrypsin inhibitor and the Kunitz anti trypsin factor. Protease inhibitors reduce availability of protein by reducing digestion and absorption. In an attempt to neutralize the harmful effects of protease inhibitors the pancreas attempt to increase protease secretion that results in the hypertrophy of the pancreas. This reaction is harmful for health.

   Protease inhibitors are also heat labile proteins and destroyed by heat treatment of feeds. Both dry and moist heat treatments are effective. The potent sources of protease inhibitor are seeds of soyabean, cluster bean, jaek bean and many leguminous seeds.

   Protease inhibitor reduces digestibility of protein causing deficiency of amino acids resulting in reduced growths and egg production. Requirement of methionine is increased due to endogenous losses.

   Common methods of destruction of protease inhibitor in feed ingredient are toasting, autoclaving or pressure flowing hot steam through the feeds.

2. **Phenolic compounds in feeds:** Two common phenotic compounds in feeds of plant origin are gossypol and tannins.

   a. **Gossypol:** It is a toxic phenolic compound of yellow colour found in cotton seeds. The content varies in different varieties and may range from 0.03 to 2.0 % in dry matter. It is an aldehyde in chemical property. Gossypol is antioxidant and inhibits polymerization reactions.

      Total gossypol in cotton seed/cotton seed cake is present in bound and free forms and the later causes toxic effects. The proportion of free gossypol is greater but variable and may be upto 70% of the total gossypol. The concentration of gossypol in cotton seed meals depends on the methods used for oil extraction. Free gossypol content in hydraulic pressed, screw pressed, solvent extracted and pre-pressed, solvent extracted cotton seed meals may be in range of 0.4 – 1.0 %, 0.2 -0.5%, 0.2 to 0.7% and 0.2 -0.3 % on dry matter basis respectively. During the process of oil extraction greater proportion of free gossypol binds with protein to form almost undegradable protein complex. This has negative effect on protein digestibility and lysine availability.

Incorporation of cotton seed meals for protein supply in composite poultry feeds should not be more than 10% to limit the concentration of free gossypol below 0.1 % which is considered the tolerance level for poultry. For the neutralization of higher gossypol content in compounded feeds ferrous sulphate can be mixed. The amount of ferrous sulphate may be upto 4 times of gossypol.

Concentration of more than 0.1% gossypol produces adverse effects on the performance and health of poultry birds. Prolonged use of cotton seed meal in the diets of poultry may be fatal. The clinical symptoms in growing chicks are loss of appetite, laboured breathing, cardiac disorders and decreased growth rate. In laying hens even small quantity causes greenish discolouration of egg yolk. Fall in oxygen transportation capacity of red blood cells results in hypoxia of tissues and associated pathological changes.

b. **Tannins:** These are complex polyphenolic compounds and constituents of many feeds of plant origin i.e. seeds, cereal grains and extractions. The two main forms of tannins in feeds are the condensed tannins and hydrolysable tannins. Small quantity of tannins may be beneficial but considerable concentration is antinutritional because tannins bind proteins and carbohydrates and render them indigestible. These also inhibit the activities of digestive enzymes (proteases, amylase and lipase etc.). More than 0.5% tannins in the composite diets of poultry reduces digestibility of proteins and carbohydrates and rate of decrease is more or less proportional to concentration of tannins but beyond 4% level may be even fatal due to drastic fall in feed intake and availability of nutrients from such feeds. Lower tannins content in feeds decreases availability of nutrients resulting in lowered growth rate in chicks and fall in egg production of layers.

Tannins containing common feeds of poultry are sorghum, milo, rape-seed-mustard seed cake and lucerne leaf meal. Non conventional feeds like deoiled sal seed meal, mangoseed kernel, tamarind seed, mahua seed kernel cake are rich in tannins. Tannins content in these feeds are very high and requires laborious and expansive technologies for nutritional improvement. Therefore, use of non-conventional feeds containing more than 2-3% tannins can not be recommended. Total concentration of tannins should not exceed 0.5% of dry matter.

3. **Glycosides:** These are complex carbohydrates which may be ester or condensation products containing alcohol, phenol, sugars and oligosaccharides. These are generally anti-nutritional for the animals. The glycosides present naturally in the feeds of plant sources are cynogens, saponins, glucosinolates and oestrogenic substances.

   a. **Cyanogenic glucosides:** Intact cyanogenic glucosides are not deleterious. Rupture of cells during grinding in gizzard releases glucosidase enzyme that hydrolyze cyanogenic glucosides and highly toxic hydrocyanic acid (HCN) is released.

      HCN is highly toxic and inhibits the cytochrome oxidase which regulates enzyme action responsible for the cell respiration. The HCN is a fatal toxin and death is spontaneous in few seconds

   b. **Saponins:** These are toxic glycosides that decrease surface tension and forms foam with proteins and sterols. Saponins decrease the availability of cholesterol by forming complex compounds which are eliminated in excreta. Level of cholesterol in blood and liver falls. Lguminous leaf meals prepared from lucerne, berseem and other clovers contain substantial amount of saponin and incorporation of high level (say 20% or more) or higher than 0.37% saponin content from leaf meals and other sources depressed body weight gain in chick. In contact with blood saponins cause haemolysis and change cell permeability. Respiration rate is increased in affected birds. Egg production decreases in laying hens.

   c. **Phytoestrogens:** The phytoestrogen more concerned with poultry nutrition is an isoflavone called genistein present in soyabean seeds. The factor is active in raw soyabeans and inactivated on heat treatment. Genistein enhanced absorption and deposition of excess amounts of several minerals like calcium, phosphorus and manganese in bones, and zinc content in liver and bones. Phytoestrogens are fat soluble substances and residue left in soyabean meal after solvent extraction is inadequate to have any appreciable effect on metabolism. Growth is depressed and feed utilization efficiency is reduced.

   d. **Glucosinolate or thioglucosinolate or goitrogens:** These are complex organic compounds responsible for reducing the availability of iodine and development of goiter for compensating the requirement of thyroid hormones. Glycosinolates or thioglucosinolates present in rape seed-mustard seed meal are hydrolysed to yield goitrogenic compounds, 2-OH-3 butanyl isothiocynate and 5-Vinyloxazolidinine-2-thione goitroin. These supperess iodine uptake resulting in goiter development. The other harmful effects are poor growth, low feed conversion efficiency, decreased egg production and haemorrhages and fatty changes in liver.

4. **Erucic acid:** It is a polyenoic unsaturated fatty acid responsible for pungent taste of rapeseed –mustard seed oil. Nominal residue is present in solvent extracted meals. Poultry feeds containing less than 0.5% erucic acid has no adverse effect on voluntary feed intake and performances of birds. The harmful effect has almost removed due to extensive production of solvent extracted cakes. Use of high proportion of expeller pressed rapeseed mustard seed cake in poultry feeds will result in reduced voluntary feed intake, poor growth response and decreased feed conversion efficiency. Now in many countries low erucic acid or non erucic acid rape-mustard seeds (e.g. canola seed) have been developed by genetic manipulations.

5. **Phytates (Salts of phytic acid):** These are organic compounds of high molecular weight formed from the esterification of a cyclic alcohol inositol with six phosphate molecules. In addition to phosphorous other dietary essential minerals also bind the complex phytate (Phytin) molecules which are constituents of plants. The concentration of phytates is generally high in protein rich feeds of plant origin. Simple stomached animals including poultry do not produce sufficient quantity of enzyme phyatse, necessary for the liberation of phosphorus and other minerals chelated in the feed. The other minerals in phytate molecules are calcium, magnesium and trace elements manganese and zinc. The balancing of these minerals in poultry feeds requires special attention for ensuring supply of these minerals. The problem is resolved by either increasing the concentration of these minerals by mixing inorganic salts of higher availability or mixing adequate quantity of exogenous phytase enzyme in compounded feeds. Six molecules of phosphate are chelated with the phytic acid and calculation of phosphorus supply only on the basis of chemical composition is highly misguiding and results in phosphorus deficiency although concentration in feeds is adequate. Phytates are also responsible for decreasing the availability of amino acids and carbohydrates.

6. **Non starch polysaccharides (NSPs):** These are constituents of feeds of plant origin. The non starch polysaccharides are cellulose and hemicellulose present in bran and these are not digested in the birds due to non-availability of enzymes required for hydrolysis into digestible and absorbable forms.

Cellulose, a fibrillar polysacchardie is a polymer of glucose molecules united by 1, 4-β bonds. Hemicelluloses and pectins constitute the matrix of cell wall complexed with lignin. Composition of hemicelluloses is complex and yet less defined. The pectin substances are polymers of galacturonic acid methyl ester. In addition to these a group of soluble NSP like b-glucans, arabinoxylon, arabinogalactons, galactomannans, galacturonans, xyloglucans and rhamnogalacturonans are harmful for the poultry. Contents of some harmful NSPs in common feeds of poultry are presented in Table 6.2.

Table 6.2: Non starch polysaccharides (NSPs) in some common poultry feeds.

Sl. No.	Non starch polysaccharide	Poultry feed ingredient
1	Arabinoxylans	Maize, wheat
2	Arabinogalactans	Mustard seed, rape seed
3	b-glucans	Barley, oats
4	Galactomannans	Cluster been seed
5	Galactoarabinans	Soyabean seed
6	Galactouranans	Soyabean seed
7	Rhamnogalactouronans	Lupin seeds

7. **Argemone (an adultrant in rape seed-mustard seed):** The prickly poppy is a weed and its seeds are easily mixed with black rape seed –mustard seed and often used for adultration of rape-mustard seed. It contains a toxic substance.

8. **Mimosine:** It is a harmful protein present in subabul (earlier and truly koobabul, Leucaena leucocephala) and binds with iodine causing iodine deficiency and decrease in level of thyroid harmones. This causes disturbances in energy and minerals (calcium and iodine) metabolism. The symptoms are reduction in growth rate of chicks, loss of body weight in adults, thinness of egg shell in layers, reproductive insufficiency or even failure and enlargement of thyroid gland (goiter). Leucaena leaf meal incorporation in poultry feeds for protein and carotenoids supply should not be recommended.

9. **Nimbidins:** Neem seed kernel cake is a rich source of protein but its feed value is significantly reduced due to presence of harmful constituents known as nimbidins. The three anti-nutritional (bitter) factors of neem seed are nimbine, nimbinin and nimbidine. Mixing of neem seed kernel cake in poultry feeds reduces voluntary feed intake resulting in decreased supply of protein, energy and other nutrients precipitating in lowered growth rate and egg production, damage of vital organs and death on prolonged feeding.

10. **Anti-vitamins:** Certain feeds contain substances that either inactivate some vitamins or reduces their potency.

    a. **Anti vitamin A:** An enzyme lipoxygenase present in raw soyabean seeds oxideses carotenoids (the percursers of vitamin A) in mixed feeds and reduces the availability of vitamin A activity resulting in occurrence of vitamin A deficiency symptoms (See vitamin A). The dietary requirement of preformed vitamin A increases.

    b. **Anti-vitamin D:** It is also present in raw soyabean seeds and increases the dietary requirement of vitamin D3 in poultry feeds for optimum functions. The deficiency symptoms are bone deformities and thinness of egg shells as observed in vitamin D deficiency.

    c. **Anti-vitamin E:** Antivitamin E factor is found in raw kidney bean (Phaseolus vulgaris). Inclusion of raw kidney bean in the diets of poultry causes development of reproductive disorders. However, problem is solved by supplementation of higher quantity of vitamin E. Probably anti-vitamin E factor in kidney bean does not interfare in the action of supplemented vitamin E which is higher due to non availability of constituent vitamin E in feed ingredients.

    d. **Anti-vitamin K:** A factor present in sweet clover (Melilotus albus) known as coumarins which are converted to discoumerol by fungi like Aspergillus species infestation during the production of leaf meal in bad weather. Dicoumarol decreases the synthesis of prothrombin in blood necessary for clotting and control of bleeding. Contamination of feeds with discoumarol increases the requirement of vitamin K in compounded poultry feeds.

    e. **Anti-vitamin B6 or Anti-pyridoxine factor:** A compound linatine present in linseed (flax seed) meal is a peptide composed of 1-amino D proline and glutamic acid. It reduces the potency of pyridoxine and produces deficiency. The deficiency symptoms are lowered appetite, reduced growth rate, encephalomalacia and chondrodystrophy. The nervous damage is exhibited by jerking movement of legs and wings. The affected chicks run uncontrolled with flapping wings and death. Some times affected birds roll on back with legs upwords. There may be paddling in some patients.

       Inappetance, loss of body weight, lowered egg production and death are common symptoms.

    f. **Anti-vitamin B_1 or Anti-thiamin factor:** An enzyme thiaminase in bracken fern (Pteridium equilinum) and raw fish meal. Thiamin activities of feeds are destroyed by mixing raw fish meal or bracken fern leaf meal in the diets of poultry.

Methods of inactivation, destruction, neutralization and extraction of anti-nutrient factors in feeds

In order to improve the nutritive and feeding values of feed ingredients various methods have been developed and standardized for practical uses by the poultry farmers and commercial feed manufacturing companies. Some useful methods are summarised in Table 6.3.

Table 6.3: Methods of inactivation, neutralization and removal of harmful factors in feed ingredients.

S.No.	Anti-nutrient in feeds	Method(s) of treatment
1	Protease inhibitors	Dry or moist heat treatments like roasting and cooking are effective treatments.
2	Haemagglutinins	Do
3	Saponins	Do
4	Cyanogens	Sun drying giving full exposure by several turning of cassava tuber chips/ small pieces.
5	Oestrogenic substances	Heat treatment
6	Glucosinolates	Solvent extracted rape seed-mustard seed meal is used below 10% to keep the level of glucosinolates below 0.6% in complete feeds.
7	Erucic acid	Genetic manipulation
8	Phytates	Phytase enzymes of fungal sources are used as effective feed supplement.
8	Gossypol	Use of cotton seed meal is avoided in poultry feeds. In shortage less than 5% is mixed in composite diets.
10.	Tannins	Small quantity of tannins containing feeds are mixed to keep the level of tannins below 0.2% in complete feeds.
11.	Argemone	Argemone contaminated feeds are not used.
12.	Mimosine	Leucaena (Subabol) leaf meal should not be used.
13.	Nimbidine	Ammoniation is claimed effective. Not yet widely used.
14.	Non-starch polysaccharides	Some commercial mixtures of poly enzymes are claimed effective
15.	Anti vitamin A	Heat treatment of soyabean
16.	Anti-vitamin D	Heat treatment of soyabean
17.	Anti-vitamin E	Kidney bean should not be fed
18.	Anti vitamin K	Sweet clover leaf meal should not be fed
19.	Anti vitamin B6	Excess vitamin B 6 should be supplemented with linseed meal
20.	Anti-vitamin B1	Thiaminase containing fish meal should be fed.

7 HARMFUL AND TOXIC FUNGI IN FOODS (MYCOTOXICOSIS AND MYCOSES)

The fungi are generally saprophytes widely distributed on the decaying organic matters, air and water. Humid tropical climate is quite congenial for the multiplication of fungi. Although different species are found in wide range of climatic conditions. so far more than ten thousand species of fungi have been identified in the nature but fortunately a small fraction of only about 50 species have been found harmful causing mycotoxicosis and mycoses in humans, animals and birds. Among the advantageous fungal species some are used as foods, viz. Mushrooms and yeast and many others are used for the preparation of fermented foods and beverages of different taste, flavour, colour and consistency like white bread, brown bread, cheese varieties, bear, wine and pickles etc. and also for the production of life saving anti-microbial drugs., i.e. antibiotics. The relatively recent use of some fungal species is growth and performance improving factors in the diets of farm animals, pets and poultry birds as probiotics.

Pathological conditions caused by fungi

Two types of pathological conditions produced by the toxins of fungi and invasion of fungi are mycotoxicosis and mycoses.

Mycotoxicoses: The harmful effects produced from the ingestion of toxins produced by pathogenic fungi is known as mycotoxicosis. The causes of disease and mortality are the level of mycotoxins ingested at a time beyond the threshold of tolerance.

Types of mycotoxins: One or more types of mycotoxins are produced by different species of pathogenic fungi. These are:

1. Hepatotoxins causing damage of liver and occurrence of associated diseases.
2. Nephrotoxins cause damage of kidneys and produce disease due to renal dysfunction and damage.

3. Vomitoxins; The harmful fungal metabolites causing stimulation of vomiting centre.
4. Neuro-muscular toxins causing degeneration of nerve fibres resulting in paralysis and dysfunction of nerves and muscles.

Mycoses

The pathological conditions produced by the invasion and proliferation of fungi in the body is known as mycoses. These may be due to increased quantity of fungus as well as the metabolites secreted by the fungus. Incidences of cutaneous and sub cutaneous mycoses are rare in poultry but incidences of systemic mycoses are observed.

Mycotoxins

Different mycotoxins identified in different kinds of foods and feeds are listed in Table 7.1. All these mycotoxins may not have direct harmful effect on the health and production of poultry but indirect harmful effects may not be ruled out. Since research on mycotoxicosis is continuing on poultry it would be advantageous to know different kinds of mycotoxins present in foods and feeds.

Table 7.1: List of mycotoxins identified in different foods and feeds.

Sl.	Mycotoxin	Sl.	Mycotoxin
1.	Aflatoxins B1, B2, G_1, G_2 and M (Modified form in milk)	15.	Microfenolic acid
2.	Alternariol	16.	Ochratoxins A, B
3.	Citrinine	17.	Patulin
4.	Cyclochanasin	18.	Pencillic acid
5.	Deoxynivalenol	19.	P.R. toxins
6.	Diacetoxyscirphenol	20.	Roquefortine A, B, C
7.	Ergot alkaloid	21.	Sporidesmin
8.	Fumiclavins A&C	22.	Sterigmatocystin
9.	Fuminosin	23.	T2 toxins
10.	Fumitoxins A, B, C	24.	Tenuazonic acid
11.	Gliotoxin	25.	Tremorgins
12.	H.T. toxins	26.	Tricothecene
13.	Kozic acid	27.	Zearelenone
14.	Lolitrem alkoliod Lusteoskyrin Alimentary toxic Aleukia (ATA)		

Important Mycotoxins and moulds of Foods/ Feeds and susceptible poultry

Salient informations on some of the mycotoxins, moulds and fungal source, foods/ feeds spoiled and poultry species affected on consumption are presented in Table 7.2.

Table 7.2: Mycotxins, fungi, foods/ feeds and susceptible poultry birds.

Mycotoxins	Fungi/ Mould	Foods/ Feeds	Susceptible poultry
1. Aflatoxins B1, B2, G1, G2	Aspergillus flavus A. perasiticum Less important are, A. bombysis A. nomius A. ochraceoroseus A. peudotamari	Cereal grains (maize, sorghum, barley, oats, wheat, seeds and meal of groundnut, cottonseed, soyabean, sesame, cassava, sweet potatoes etc.	Young birds, chicks, quail, pheasants, turkey
2. Ochhratoxins A and B	Aspergillus allicaceus A. melleus A. ostianus A. petrakii A. Scler otiorum A. sulphureus Penicillium commune P. cyclopium P. purpurescens P. viridicatum	All above loisted ingredietns	Chicks, Ducklings
3. Patulin	Aspergillus clavatus A. giganteus A. terreus Pencillium claviforme P. cyclopum P. equinum P. expansum P. leucopus P. melinii P. patulum P. urticae Paecilomyces spp.	Apple and other fruits and their non fermented products.	Young bird chicks, quail, wild birds
4. Pencillic acid	Aspergillus melleus A. ochraceus A. quercinus	Dried beans and seeds	Many birds

[Table Contd.

Contd. Table]

Mycotoxins	Fungi/ Mould	Foods/ Feeds	Susceptible poultry
	A. sulphureus Penicillium bearneuse P. cyclopicum P. madriti P. marteusii P. puberulum P. suavolens P. thomii		
5. Luteoskyrin Sterigmatocystin	Aspergillus nidulus A. regulosus A. versicolor Penicillium luteum	Starchy grains	Consumers of infected grains
6. Citrinin	Aspergillus oryzae, Pencillium camumberti P. citrinum	Wheat, Barley, oats, rice, maize, rye	Poultry birds
Recent species	Monoscus ruber M. purpureus	Used for production of red colour	
7. Roquefortine A, B, C	Pencillium, roqueforti	Blue chease	
8. Alimentary toxic aleukia (ATA)	Species of Alternaria, Cladosporium, Fusarium, Mucor Pencillium		
9. Ergot alkaloid	Claaviceps paspali, C. purpurea	Paspalum Pearl millet, rye	Poultry and other birds
10. Zearalenone	Fusarium crookwellense F. graminearum F. culmorum F. equisetum F. monoliformis F. oxysporium F. roseum F. tricutum	Many foods and Feeds	Poultry
11. Trichothecene (T2 toxins)	Fusarium tricincutum	Wheat, maize etc.	Poultry
12. Fusarins	Fusarium spp.		
13. Equisetin	Fasarium spp.		
14. Beauversin	Fusarium spp.		
15. Enniatins	Fusarium spp.		
16. Butenolide	Fusarium spp.		

Treatments of contaminated foods/ feeds for improvement of food value

Various methods used for the treatment of contaminated food stuffs are used to eliminate, deactivate or reduce the levels of toxins below harmful effects. All methods may not be equally effective. Therefore, specific treatment method should be used for the detoxification, where it has been standardized. Various methods used for controlling toxic effects are the use of binders, deactivators, physical separation, washing, milling and use of chemical agents.

1. **Use of additives:** Various substances are used for binding and deactivation of mycotoxins in foods/ feeds. These substances either adsorb or deactivate the mycotoxins to make the food nontoxic for poultry feeding. Some of the known additives are montimorillonite, bentonite clay, activated charcoal, butylated hydroxyl anisole (BHA) and butylated hydroxitoluene (BHT). Some other binders, kaolin, hydrated sodium-calcium aluminosilicate (HSCAS), zeolites and glucomannan (Non starch oligopolysaccharide) are also used. A herbal product made of extracts of Acacia catechu, Andrographis paniculota and Phylanthus niruri in a non toxic misible base has been found effective for removing or reducing the toxic effects of mycotoxins.

   The efficacy of feed additives used for the control or reduction of harmful effects of mycotoxins on the body is assessed on the basis of following criteria:

   (a) Mycotoxins adsorption potential of the binder.

   (b) Requirement should be quite small, i.e. normally less than 1 percent of the foods.

   (c) Should be effective in wide pH range of digestive system.

   (d) Additives should be non toxic and eco friendly.

2. **Use of mycotoxins deactivators:** These are enzymes (esterase and epoxidase), yeast (Trichosporon mycotoxinvorans) or bacterial cultures (Fubacterium BBSH 797).

3. **Other methods of mycotoxins removal from the foods and feed:** The effectiveness of these methods are highly variable and depends on the standardization and application of the techniqes. These are physical separation of damaged grains and foods, washing, milling, heat treatment, radiation, solvent extraction etc.

Factors affecting the harmful effects of mycotoxins:

1. Ulthriftness due to under feeding and deficiencies of energy, protein, vitamins etc.
2. Synergistic effect of other infections in the body
3. Unhygeinic housing.
4. Dirty drinking water containing organic matters.
5. Synergistic effect of other toxins present in the body or present in the foods containing mycotoxins.
6. Quantity of toxins ingested.
7. Duration of mycotoxins intake.
8. Age, sex and health of the exposed subject.

Genesis of the term mycotoxin

The term mycotoxin was coined by a group of veterinarians investigating the cause of mortality of over 100000 turkey poults at a poultry farm near London in England during 1962. The disease was initially named x-disease but the cause could be known after the examination of feed ingredients and other things at the farm. The groundnut meal fed to the poults was found heavily contaminated with Aspergillus flavus and contained high concentration of toxic secondary metabolite. The greek word mykes or mukos was used as" Myco" and latin word "toxin" for poisonous substance to constitute the word "Mycotoxin" means the toxin produced by fungus particularly the mould (Table 7.3).

Table 7.3: Year of invention of some mycotoxins.

Sl.			
1	Aflatoxins	1962	
2	Citrunin	World war II	
3	Ochratoxins	1965	Ochratoxin A
4	Fumonisins	1988	
5	Petulin	1940s	Antibiotic (Penicillin)
6	Ergot alkaloids	1934	

Conditions for growth of toxin producing moulds

1. Hot–humid climate of tropical and sub tropical zones is most favorable.
2. Almost all kinds of natural organic matters-live, dead and decaying are substrates for moulds and other fungi.

3. Invisible organic matter in soils are also substrates for fungi.
4. Wilting and dehydration of green/wet crops favour mycotic growth. This may be observed during drought.
5. Damage of crops by insects and other factors predispose fungal contamination, multiplication and mycotoxins production.
6. Handling, shipping and storage increase fungal contamination, growth and mycotoxins production.

Effects of mycotoxins on poultry products

1. High mycotoxins content in poultry feeds may cause only one stage economic loss due to mortality of flock.
2. Gradual intake of low levels of mycotoins is more hazardous for the birds causing drop in feed intake, lowered growth, lowered egg production, prolonged sickness and death.
3. Consumption of eggs and meat from poultry birds fed mycotoxins contaminated feeds are highly hazardous for humans eating such eggs and meat. The most dangerous effect is carcinogenesis.

Prevention of mould contamination and mycotoxins production in foods/feeds

1. Treatment of seeds with antifungal chemicals before sowing.
2. Physical removal and destruction of mouldy plants and feeds, preferably by burning
3. Application/spray of fungicidal chemicals on standing crops during emergence of head and grain formation.
4. Drying of food grains, oil seeds and other foods to contain less than 12% (preferably 10%) moisture.
5. Harvesting conditions: Crops should be harvested in dry weather and should be subjected to quick drying by applying quick turning.
6. Food grains and other crops should be dried on hard floor, terpolin or thick plastic sheets.
7. Mechanical drying should be used with draft air blowing through the food grains beneath a tin shade or other suitable dry place.

8. Food grains and other edible crops should be cleaned first before drying and again before storage to remove damaged grains, infected grains and spoiled materials.
9. Storage places, utensils, sacs, godowns etc. should be clean and dry. Fumigation of godowns and hard wares should be done before storage of food grains and feedstuffs.
10. Storage places should be well ventilated but entrance of moist air should be prevented.
11. Storage should be free from insects, birds and rodents. These may be carriers of fungal spores.
12. Strict hygienic conditions should be maintained at the poultry farm. Daily cleaning of cages, feeder, waterer and feed handing appliances is necessary. These should be properly disinfected at short intervals.

Testing before procurement

Some common feeds like maize grain, sorghum, groundnut cake, cotton seed cake, sunflower cake, barley, rice bran, rice polish and wheat bran require more attention for examination at the time of procurement. Detection of the presence as well as type and concentration of different mycotoxins is important for the selection of feedstuffs and determination of their ratio in the compounded complete feeds to maintain the levels of mycotoxins (if present) below the tolerance level.

For the quantitative estimation of mycotoxins sensitive instruments are required. Some of the methods used are thin layer chromatography (TLC), high pressure liquid chromatography (HPLC), gas – liquid chromatography (GLC) and immoassys.

Sampling of foods/ feeds for analysis

It is the first most important step for the correct detection of mycotoxins in the foods/ feeds and other samples. Sampling is a tedious exercise from big lots of materials stored in bins, sacs and silo. For such situations physical examination should be done thoroughly and solid spots, odour emitting spots and discolored spots should be identified on the piled sacks. The nutritionist at the factory should use his experience for sampling.

Since poultry farmers purchase compounded feeds in lots, it will be easier to take representative samples for analyses. Results may be available with in 3 days and farmers should arrange the frequency of feed procurement in a manner to get new feed stock 7-10 days before the exhaust of feed in use.

Following measures should be taken in collection, packing, sealing and delivery of feed/food samples for analyses:

1. Sample should be representative and should represent sampling from different rows, columns and strata.
2. At random samples should be taken out and thoroughly mixed before sub-sampling.
3. If sample is not dry, it should be dried at 80-90°C for 3-4 hours for the estimation of mycotoxins but only at 55-60°C for the culture of sample for the detection of mould by culture.
4. The mixed samples collected from the lot are now filled in paper bags or cloth bags, tied and sealed. Details of feeds/ foods along with data of procurement and date of sampling and the details of factory to which it belong should be written on the bags or should accompany the bag(s).
5. Three bags of each lot is prepared. One is sent to laboratory for analyses and 2 bags are kept in safe custody for use at the time of dispute, if any.
6. Plastic or moisture proof materials should not be used for sample transport, because moisture produced during transpiration of grains or evaporation from other foods may condense and wet some amount of sample. This may cause activation of fungal spores, if present in the sample. However, plastic bags can be used, if transportation is in frozen form or storage is at refrigeration temperature.

Season: mould relationship

Although moulds are active in all seasons but relative activity is influenced by atmospheric temperature and humidity. This ultimately affect the mycotoxins concentration in feedstuffs as shown in table 7.4.

Table 7.4: Effect of season on mycotoxins production.

Season	Mycotoxins in foods/ feeds
Wet winter	Ochratoxins, Fuminosins, T2 toxin, Vomitoxin, Diactoxyscirpenol (DAS), zearalenone
Hot-humid (Wet summer)	Almost all kinds of mycotoxins, viz., Aflatoxins, ochratoxins, by Aspergillus, infection only, fumonisins, Ergot alkaloids.
Dry hot	Ergot alkaloids on pearl millet.

Effect of mycotoxins on immune system

Many mycotoxins have been found deleterious for the immune system in poultry and other animals. Immuno suppression effect of some mycotoxins reduces disease resistance ability of birds and increases susceptibility for the infectious diseases. Immuno suppression damage is maximum by the aflatoxins followed by vomitoxins, T-2 toxin, ochratoxins and fumonisins in descending order.

The main organs of humoral and cellular immunity in poultry are the bursa of fabricius, thymus and spleen. The other organs contributing to immunity are the caecal tonsils and bone marrow. The T-cells produced in the thymus are responsible for the development of cell mediated immune system while B-cells produced in the bursa of fabricius and bone marrow are responsible for the development of humoral immunity by synthesizing antibodies or immunoglobins (Ig M and IgA).

The immunosuppression effect is highest by Aflatoxins and had damaging effects on both cell mediated as well as humoral immune systems. The cell-mediated immune systems are damaged by low level of aflatoxins in the body whereas high concentration suppresses humoral immunity systems. The trichothecenes are next to aflatoxins for damaging immune systems in chicken. These toxins damage the immunogenic cells on the sites of origin in spleen, thymus, lymph nodes, bone marrow and intestinal mucosa. The ochratoxins damage results in the atrophy of thymus and lowering of both humoral and cell-mediated immune action. The levels of circulating immunoglobulins and phagocytes fall and rate of fall depends on the levels of ochratoxins ingestion and duration of intake by birds.

8

PHYSICAL QUALITY ASSAY OF FEEDS

Quality of feed ingredients is one of the most important factor having significant effect on the health and production of poultry. Therefore, utmost care should be taken for the supply of wholesome feeds. The quality maintenance is required at all stages starting from procurement of ingredients to processing, compounding, supply and storage before feeding. Three methods of feed quality evaluation in vogue are the physical, chemical and biological methods. The later two are the laboratory methods whereas first one is used for the procurement of feed ingredients for compounded feed production.

The methods of physical quality assessment should be simple, easily detectable of differences in different samples of same feed ingredient and generally should not require costly instruments. However, some times use of wide field microscope may be required, specially for the detection of adultrations in crushed feeds and brans etc. Physical methods of feed evaluation are crude processes but important for the procurement of ingredients. The importance of physical evaluation of feed ingredients is higher in tropical countries where grading of grains and storage of food grains, pulses and oil seeds are highly irregular and at many places get exposer to environmental factors like heat, moisture, dust, bacteria, fungi and insets etc.

Physical properties of feeds

The knowledge of physical properties of feed ingredients and compounded feeds is essential for the differentiation and grading of feeds. The physical characteristics of feeds are assessed on the basis of soundness (shape, size and density) of food grains, moisture content or dampness of the feeds, colour of the feeds, taste, odour, presence of insects and fungi and cake formation in stacked feeds.

Physical methods of feed assay

The first step of quality control is the physical assay of feed ingredients and compounded feeds. The techniques of physical assay of feeds should be easy and feasible without much instrumentation and can be done by following methods.

1. **Visual inspection:** Satisfactory assortment of many feeds particularly the whole grains, seeds and kernels is quite simple for the farmer and traders of food grains, pulses and oil seeds. Visual inspection of feeds color, shape, size and detection of adulteration with physically comparable materials.

   (i) Color: The color of feeds should be characteristic and generally uniform. Almost all undamaged food grains, leguminous seeds (pulses) and oil seeds appear bright. Characteristic color of yellow maize is dark yellow to radish yellow. The color of sorghum may be white, dull or yellowish white and red in different varieties. Mixing of dust and extensive application of insecticides and pesticides changes the natural color of undamaged food grains. The lot of such feed is dusty and dull in appearance.

   Dark color sorghum commonly known as Milo contains high level of naturally occurring tannins and use of brownish to dark red sorghum in preparation of poultry feeds should be avoided. Digestibility and availability of nutrients from milo is much lower than the white sorghum, which is comparable with white maize. Similarly there is conspicuous difference in the color of rape seed/ mustard seep, viz. black, yellow, brown and brownish red. The color of rape seed/ mustard seed cake corresponds closely with the seed color. Howere, nutritive value of cakes from different color seeds is comparable for poultry.

   (ii) Size: The size of food grains depends on variety and it should be uniform for a variety of cereal grain, leguminous seed and oil seed. With reduction in the size of grains the proportion of less digestible outer coat (bran, husk and hulls) increases and digestibility and metabolizable energy values are decreased. Variation in the size of grains is generally due to mixing of two or more varieties. A simple method of grading maize grain on the basis of size has been suggested. A lot of maize grain containing cut, broken, touched, heat damaged and discolored maize grains less than 5%, 5.1 to 10% 10.1 to 15% and 15.1-29% may be classified in grades 1,2,3,4, and 5 respectively. All lots of maize containing more than 20.1% deformed grains should mot be accepted. However, acceptance of more than 10% deformed grains in the lots of maize for use in poultry feeds needs reconsideration considering all aspects of damage. It may be more logical to remove the touched, discolored and heat damaged grains from grades 2 and 3. There is no harm in considering cut and broken maize grains for poultry feeding.

(iii) **Shape:** The shape of food grains including legumes and oil seeds should be characteristic for the species and variety. The normal shape is usually changed by water soaking and dampness. Drying of water soaked food grains brings visible change in the normal shape of food grains. The change is mostly associated with dullness of color. Such food grains are seen in flood affected and heavy rainfall areas. The washed grains extracted from the dung of bovines used for thrashing without putting muzzle are also abnormal in shape.

(iv) **Detection of adulterants:** Some non-edible or cheap ingredients of low feeding value are frequently used for the adulteration of feed ingredients. Seeds of argemone are more or less similar to black rape seed-mustard seed and it is often difficult to identify by inspection. The surface of seeds of rape/ mustard is smooth and bright while that of argenone is rough and dull. Similarly in an adultrated wheat bran with ground rice hulls, the later is detected by its hardness and roughness. However, a recent technique of dipping rice husk in dilute acid before drying and grinding is causing hindrance in physical detection.

(v) **Presence of insects:** Considerable damage of grains is quite common in tropical environment. The most common insect is tribolium which eats endosperm living behind the empty fibrous coat. It further contaminates the feed by excreting uric acid.

2. **Touch and hand manipulation of lots:** Dampness or sensible moisture content can be detected by touch. By pushing (penetrating) hand inside the lot of food grains, approximate extent of dust may be detected on the basis of the layer of dust formed on the surface of hand. There may be clumps formation due to improper storage of whole grains and processed feeds. High moisture content (more than 12%) in food grains, immediate packing of hot solvent extracted oil seed meals, high moisture hot pellets and other processed feeds produce clumps. On taking a handful broken rice, rice bran and rice indicates significant amount of oil in the lot and on storage it will be spoiled by oxidative rancidity.

In case of grains stored for long time in large godowns penetrating of hand with fist for taking out handful sample, if there is spontaneous itching sensation, coughing and sneezing, the formation of fungal toxins and spores may be suspected. It was experienced by the author during visit to different warehouses of Food Corporation of India for the purchase of damaged wheat for livestock feeding. While handling such suspected lots of food grains and other feeds glove and mask should be used for sample taking.

3. **Odour (Smell):** All food grains, milling products and by products and other feeds emit characteristic odour of variable intensity. The odour of cereal grains and their milling by products are generally mild and sweet, which becomes unpleasant and repelling on development of rancidity. Similarly any abnormal change in characteristic odour of a feed should be given special attention for quality assessment. Following guidelines may be used for suspecting spoilage and reasons for spoilage of the feeds. (Table 8.1)

Table 8.1: Type of off smell (flavor) and reasons for deterioration of characteristic odour.

Sl. No.	Type of odour	Suspected causes
1.	Sour and repulsive	Mouldy growth, damage by insect.
2.	Musty smell	Fungal growth and insect boring.
3.	Like petroleum products	Use of chemical insecticides, pesticides and fungicides in heavy doses beyond the recommended doses.
4.	Tallowy due to rancidity	Oxidative changes in fat content of feeds due to high oil content and lack of anti-oxidant factors or additives.
5.	Heat damaged odour	Depending on the degree of heat damaging effects, the smell may be tolerable. Experienced person can even detect the feed burnt.
6.	Leathery smell	Contamination of meat meal and meat-cum-bone meal with leather meal.
7.	Pungent smell of ammonia	Emitance of ammonia indicates adultration of fish meal and other feeds for increasing the content of crude protein (Nx6.25) in the feed. Urea is also used in adulteration of groundnut cake with rice polish.

4. **Taste:** Any change in characteristic taste of feed ingredient or processed products of feed ingredients should be doubted for spoilage or contamination. Bitterness in oil seed meals of groundnut, soyabean, sunflower and coconut meal indicates presence of mycotoxins due to fungal infestation. Bitter taste also indicates adulteration with mahua seed kernel cake and safflower seed cake.

5. **Sound:** Detection of sound of a feed ingredient or compounded feed depends on experience. Sound of properly dried maize, sorghum, wheat and pearl millet on pouring from sacks resembles the sound of spilling coins. The sound of barley and oats is mild than the sound produced by maize, sorghum and wheat.

6. **Effect of hammering:** On hammering properly dry grains and feeds are broken in pieces whereas that containing higher moisture flatten. Some experienced testers make assessment of dryness by cracking between teeth.

Some physical spot tests for physical evaluation of feeds

Generally four tests are used for the physical evaluation of feed ingredients. These tests are more useful for the identification of adulterants in feed ingredients at the time of purchase. These tests are sieving, weighing, water mixing and wind blowing for separation of foreign materials used for adulteration.

a. **Sieving:** A representative sample of feeds to be tested is sieved through a common sieve for the separation of other substances/ adulterants.

b. **Weighing:** This method is more specific for the selection of grains of optimum quality. In this process random samples of grains are collected and random 100 grains/seeds are weighed and compared with the standard weight determined and standardized by a competent authority responsible for the quality control of feedstuffs.

This test is also used for the detection of cut, broken and damaged grains by weighing a representative sample of a food grain free from foreign substances and adulterants. As per USA test, weight of a known volume (Bushel) is weighed. Example of USA for maize, barley, oats and sorghum is presented in Table 8.2.

Table 8.2: US grade of common food grains (kg/bushel).

Food grain & grade	Min weight per bushel		Percent of damaged/ foreign material					
			Sound grain	Cut grain	Heat damaged	Foreign materials	Broken kernel	Thin kernel
	Lb	Kg						
Maize	Zea mays							
1	56	25		3	0.1	–	2	–
2	54	24		5	0.2	–	3	–
3	52	23		7	0.5	–	4	–
4	49	22		10	1.0	–	5	–
5	46	21		15	3.0	–	7	–

[Table Contd.

Contd. Table]

Food grain & grade	Min weight bushe	Percent of damaged/ foreign material					
		Sound grain	Cut grain	Heat damaged	Foreign materials	Broken kernel	Thin kernel
Maize	**Zea mays**						
Sorghum	(Sorghum vulgare)						
1	56		2	0.2	–	4	
2	55		5	0.5	–	8	
3	53		10	1.0	–	12	
4	51		15	3.0	–	15	
Barley	(Hordeum vulgare)						
1	47	97	2	0.2	1	4	10
2	45	94	4	0.3	2	8	15
3	43	90	6	0.5	3	12	25
4	40	85	8	1.0	4	18	35
5	36	75	10	3.0	5	28	75
Qats	(Avena sativa)						
1	36	97	–	0.1		2	2**
2	33	94	–	0.3		3	3
3	30	90	–	1.0		4	5
4	27	80	–	3.0		5	10

Note: *Includes foreign materials, broken grains and grains. *Wild oat grains.

In India test weight for maize kernel has been worked out which is widely accepted. Weight of 100 liters maize grain should be 725 to 775 kg or 1 liter should be 725 to 775g. Acceptable standard values for other food grains are probably not available. Bulk density of common feeds is presented in Table 8.3.

Table 8.3: Bulk density of common feeds.

Sl. No.	Feeds	bulk density (g/liter)
1.	Maize	725 to 775
2.	Sorghum (jowar)	700 to 770
3.	Pearl millet (Bajra)	720 to 760

[Table Contd.

Contd. Table]

Sl. No.	Feeds	bulk density (g/liter)
4.	Small millet (Ragi etc.)	700 to 750
5.	Rice / broken rice	700 to 775
6.	Wheat	700 to 770
7.	Barley	550 to 600
8.	Soybean	520 to 570
9.	Groundnut cake	650 to 700
10.	Rape/ mustard seed cake	650 to 675
11.	Sun flower meal	500 to 530
12.	Rice bran (extracted)	350 to 400
13.	Rice polish	400 to 420
14.	Fish meal	725 to 775

Test for detection of sand and dust

A known (weighed) representative sample of test food grain is mixed with adequate water preferably in a glass cylinder of 1 or 2 liters capacity. The mixture is gently stirred 4-5 times and then allowed to settle. After few minutes 2 distinct layers are formed, if only sand is mixed. The two layers are collected in separate Petri dishes or other trays and dried at 100°±2°C for at least 6 hours. After this both constituents are weighed and percentage of sand in feed is worked out. In case of dust containing clay separation of dust may need scrubbing of grains due to formation of mud layer on grains after mixing with water. This method is more useful for on the spot qualitative assay for the presence of sand and dust in the feed. Some times a small granule of marble or white stone is adulterated with white maize, white sorghum and broken rice. Mixing is more common with broken rice.

Homogenesity

The food grains should be uniform in shape and size and that should be characteristic for the feeds. The feed should not contain foreign materials like weed seeds, other cheaper grains and damaged grains. Normally color, shape and size of the food grains are considered for the visual assay of feeds. Caking and lump formation occurs in processed feeds, ground feeds, fish meal and milk powder on bad storage and dampness in godowns.

Factors to be considered for Appraisal of poultry feeds

The common factors to be considered for visual appraisal of poultry feeds at the time of purchase of feed ingredients, processing for compounding, manufacture of compounded feeds, storage and also during feeding are summarized in table 8.4.

Table 8.4: Physical characteristics of common poultry feeds and deviations affecting quality.

Sl.no.	Feed ingredient	Factors for observation	Common adultrants
A.	**Cereal grains**		
1.	Maize	Color, shape, size	Cob pieces, dust
2.	Sorghum	Bulk density	Dust, sand
3.	Pearl millet	Damaged grains	Dust, sand
4.	Wheat	Percentage, mixture of	Dust, sand, weeds
5.	Rice	other grains, weed seeds,	White stone grits
6.	Barley	moisture, adultrants etc.	Husk, straw
7.	Oats		Husk, straw wild oats.
B.	**Cereal grains milling byproducts**		
1.	Wheat bran	Color, smell, caking, moisture, adulterants.	Ground rice hulls, saw dust, sand, dust etc.
2.	Rice polish	Color, rancidity, caking, dustiness.	ground rice hulls, sand, dust.
3.	Rice bran	-do-	-do-
C.	**Oilseed cakes (meal)**		
1.	Soybean seed	Freshness, moisture, bulk density, color,	Other seeds.
2.	Soybean meal	Rancidity, color, odour, taste etc.	Extraction with rice polish, mould growth.
3.	Groundnut cake	Freshness, bulk density, heat, clumps.	Mould growth, extraction with rice polish.
4.	Rape seed/ mustard seed cake	Freshness, moisture, clumps, heat etc.	Argemone (chemical analysis).
D.	**Animal product**		
1.	Dry fish	Sand, salt, moisture.	Sand, salt.
2.	Fish meal	Sand, salt, moisture, guts, scale, fins etc.	Sand, salt, other marime materials.
3.	Meat maeal	Color, odour, sand.	Leather meal, sand.
4.	Meat-cum bone meal	Color, odour.	Bones, sand

[Table Contd.

Contd. Table]

Sl.no.	Feed ingredient	Factors for observation	Common adultrants
5.	Skim milk powder	Color, odour, clumps, dust.	Over heating dust.
E	**Mineral supplements**		
1.	Phosphates of calcium	Moisture, clumps	Sand, fluorine (chemical test) clay, ash
2.	Bone meal	Bone pieces, sand, clay dust	Sand, clay dust, ash
3.	Mineral mixture	Moisture, color, flow, clumps	Sand, clay dust, ash
F	**Vitamin supplements**		
	These are almost proprietary products and leaflets or instruction for composition, expiry date and method of use should be read thoroughly before use.		

9 THE DIGESTIVE SYSTEM DIGESTION, ABSORPTION AND METABOLISM

The digestive system of avian species is simple and similar to swine with some significant difference in some of the segments. There is also difference in the segments of digestive system of various avian species differing in food habits. The digestive system of fowl has been taken as the representative of avian species. Some of the apparent differences in the anatomy of the digestive system of fowl from the digestive system of simple stomached mammal like pig are presented as follows:

1. The lips, cheaks (buckae) and teeth are absent is birds.
2. These have been replaced by a hard beak of various shape and size in different species.
3. There is a large size oval diverticulum nearly in the middle of the esophagus. This is called crop and used for the storage of food for soaking and softening. The surface contains secretary glands. The secretion in the crop of pigeon is called crop milk.
4. True stomach or proventriculus is elongated and glandular.
5. This is followed by a modified muscular stomach known as gizzard. The thickness and strength of muscles are higher in grain eating birds.
6. The soft palate is absent in most of the avian species.
7. The hard palate communicates with the nasal cavity.
8. There is no sharp demarcation between the mouth and pharynx. The mouth of birds cam be called oral cavity but not buccal cavity.
9. The proventriculus is the true glandular stomach. It also works as storage compartment in avian species devoid of a crop and many aquatic and semi-aquatic birds eating fish, tadpole, spawn and other aquatic animals.

10. The gizzard of carnivorous birds is normally lighter than the gizzard of grain and seed pickers. In the gizzard of most of the carnivorous birds the two pairs of muscles, namely musculi intermedii and the musculi laterales are absent. Some of the common exaples are heron, hawk and owl.
11. The small intestine posterior to duodenum is not marked into jejunum and ileum. There may be a mark of yolk sac, known as Meckel's diverticulum in the middle portion of the small intestine.
12. The length and volume of small intestine and caeca are much higher in herbivorous birds than the carnivorous birds.
13. Microbial digestion system in the caeca is more developed in herbivorous birds.
14. Bruner's glands are absent in fowl but in some other avian species homologous tubular glands are found.
15. The villi of small intestine contain a net work of blood capillaries but lacteals are absent.
16. Well formed functional caeca are found only in grain pickers and herbivorous birds like fowl, guinea fowl etc.
17. The relative volume of spindle shaped crop of goose is much higher than the crop of chicken.
18. The thin and thick muscle of gizzard of goose produces more than double grinding effect than the gizzard of chicken, and it is also higher in duck.
19. The large intestine is short and clearly not separated in colon and rectum.
20. The cloaca is the terminal segment of digestive system differentiated into three areas. This is a common passage for the excretion of mixed urine and faeces together, and also for copulation and egg laying.
21. Soft palate is present in pigeon and its relatives.

The length of alimentary canal of Fowl and some other birds

The alimentary canal or digestive system of birds is a muscular canal with lumen lined with different types of epithelium and covered in tunica serosa. The organ is differentiated in to specialized compartments for performing specific functions like prehension, collection or transient storage, moistening, enzyme treatment, grinding, digesting absorption of digested nutrients and excretion. The compartments of the digestive system of fowl are the oral cavity, pharyrnx, esophagus or gullet, crop, proventiculus, gizzard, duodenum, small intestine, caeca, colon, rectum and cloaca. The accessory glands of digestion are the salivary glands, liver and pancreas. The cloaca or vent is annulated in three subsequent sections, i.e, coprodium, urodium and proctodium controlled by a sphincter at the exit. The cloaca is a common passage for the exertion of faeces and urine together, laying of eggs, and deposition and transportation of sperm in the reproductive tract. Average length of different components of the alimentary canal of fowl (Gallus domesticus) and guinea fowl (Numida meleagris) are presented in Table 9.1.

Table 9.1: Average length (cm) of different compartments of alimentary canal of adult poultry birds.

Compartments of alimentary canal	Chicken or fowl	Guinea fowl
Precrop oesophagus	20	18
Oesophagus with crop	35	27
Proventriculus	15	5
gizzard (ventriculus)	6	6
Duodenum	20	18
Small intestine	120	125
Caeca	17	20
Colon + rectum	11	10
Total length (cm)	224	229

The relative capacity of digestive system of duck is higher than the chicken and that of goose is higher than the duck. The larger segments are crop, small intestine and caeca for holding and digestion of herbaceous foods because intake of aquatic plants and other forages is much higher in goose followed by duck.

The gizzard is made of two pairs of strong muscles known as muscnli intermedii and mucnli laterales which are also called thin muscle pair and thick muscle pair respectively. These muscles are absent in most of the carnivorous birds like heron, hawk etc.

The size of proventriculus is much higher in goose and duck consuming higher proportion of herbages. The proventriculus is also muscular than that of chicken. The muscles of gizzard are also much stronger than the gizzard of chicken due to which pressure developed in gizzard during the process of digestion is about 275, 180 and 125 mm Hg (mercury) in goose, duck and chicken respectively.

The digestive system and associated glands of the chicken

Mouth or oral cavity: The front horny part is known as beak. It is formed of upper horny part, lower horny part and hard palate. Both parts of beak are mobile and form the main organ of prehension. Soft palate is absent. The tongue is narrow and pointed interiorly. A slit in hard palate connects nostril for the breathing. The posterior part of the tongue is rough. The salivary glands drain in the mouth. The mouth joins oesophagus through the short pharynx.

The oesophagus is a long muscnlo epithelial tube which is modified in a capacious diverticnlum in posterior half known as crop. The crop is a transit store for food. The oesophagus opens into proventriculus or glandular stomach which continues as a strong muscnlar gizzard. The grinding force of gizzard is very strong in grain pickers and grazing birds and grits are frequently ingested by grazing birds for increasing the grinding efficiency. Earlier it was a common practice to offer grits to poultry for voluntary intake but now it is not used in farm birds.

The gizzard opens into duodenum which is drainage site for liver and pancreas. This continues as small intestine of long length. The digesta of small intestine drains into caeca. The caeca harbour numerous useful micro organisms that produce fibrolytic enzymes for the digestion of fibrous carbohydrates to produce volatile fatty acids. The digesta enter into large intestine (colon and rectum) and then into the cloaca to be excreted out along with urine deposit. (Fig. 9.1).

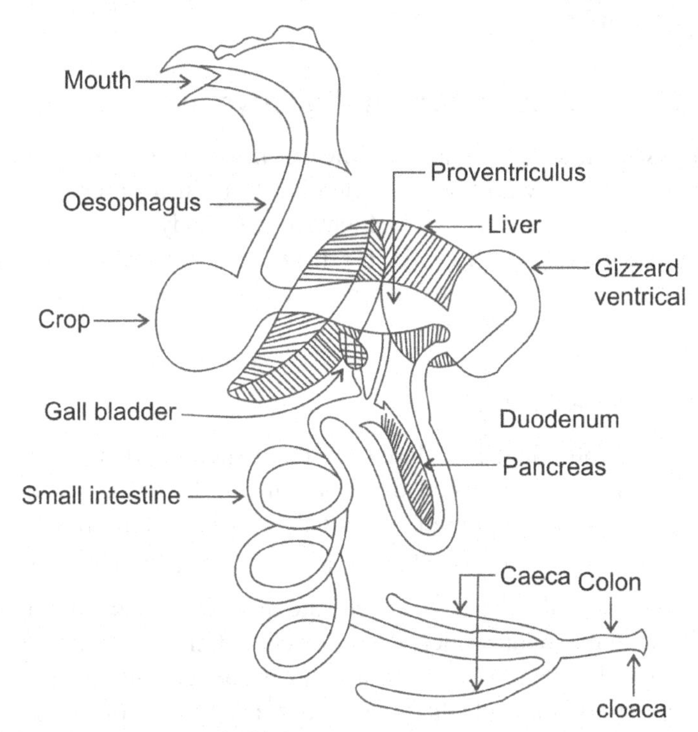

Fig. 9.1: Dugestuve system of chicken

Appetite and food intake

Like mammals the center of appetite regulates feed intake in birds. The emptying of stomach and lower tract sends message to appetite centre in hypothalamus for response of food intake. The other factors responsible for affecting voluntary feed intake in birds are as follows:

(i) Factors decreasing the food intake are low palatability, higher concentration of high energy feeds, higher concentration of protein, elevated ambient temperature and low supply of drinking water.

(ii) Factors increasing the food intake are high palatability, softness, lower ambient temperature, molting, active phase of growth and laying and dilution of energy and/ or protein in the diet. Available energy content in the diet has more influence on voluntary feed intake than the protein content. Experiments have shown the ability of near optimum protein and methionine selection from the diets of variable density of protein offered in cafeteria system. Adult chicken prefer the diet with 15-17% protein than the diets of 8-10% or 22-24% protein content.

Damage of hypothalamus had adverse effect on appetite. The damage of lateral part of the hypothalamus causes aphagia whereas ventromedial damage results in hyperphagia.

Motility of different segments of digestive system

Motility of digestive tract is essential for the swallowing of food and water, and their progressive passage with variable retention time in different segments for digestion of foods and absorption of the nutrients into the body for further distribution to target organs for the release of energy for different vital function and utilization of nutrients for the synthesis of tissues required for growth, reproduction and repair of wear and tear of the organs.

Deglutition or swallowing

Deglutition of feed and water is different in avian species when compared with the mammals. Some differences also occur among the avian species due to difference in the structure of mouth. Soft palate is absent in almost all fowl, guinea fowl, turkey and quail etc. but it is present in pigeon and dove.

After picking food the head is raised and neck is stretched for posterior movement of food in the mouth which is swallowed with the coordinated movement of the tongue, hyoid apparatus and larynx and propelled through pharynx into oesophagus. The stimulation of pharyngeal roof and tongue by the presence of food or fluid results in the closer of glottis and nasal slit on the hard palate.

Raising of head is not required for the deglutition of food and fluid in pigeon and dove. The process of swallowing is probably facilitated by the presence of soft palate.

Motility of oesophagus

The movement of food and fluid is facilitated by the peristalsis and the stage of gastric contraction cycle determines the diversion of ingested material into crop or direct in the proventriculus. The peristaltic movement of crop is stimulated by the emptying of proventriculus and lower parts. The process is coordinated by nervous reflex stimulated by hunger, passage of ingesta to distal part and defaecation.

Motility of proventriculus, gizzard and duodenum

The process is neuro-muscular in nature and influenced by hunger, thrust, emptying of posterior parts and defaecation. The contractions of proventriculus anteriorly and duodenum posteriorly are connected with the intrinsic neural connections with the muscles of the gizzard. This is responsible for the gastro duodenal peristalsis sequence in chicken. In this process contraction starts from the thin muscle (musculi intermedii). This is followed by 2-3 peristaltic movements of duodenum. This is followed by the contraction of thick muscle (musculi laterales) of the gizzard carried forward to peristaltic contraction to proventiculus. In the process of contraction of the thin and thick muscle pairs of the gizzard the wave moves in a counter clock wise manner in each muscle.

In this counter clock wise cyclic peristaltic contraction ingesta moves forward from the gizzard to duodenum during gastro duodenal phase of contraction. The passage of ingesta from oral cavity and crop to the proventriculus occurs during the contraction of thick muscle.

The gastro-duodenal segment of the avian alimentary tract is also associated with the two other type of anterior luminal movements. In this process a reflux of duodenal and adjacent ileal digesta into gizzard occurs. This helps in the mixing of the contents of proventriculus with that of duodenoileal digesta in the gizzard.

The normal rate of contraction of stomachs (proventriculus and gizzard) in chicken is 2-3 per minute. The average value of intra luminal pressure of the gizzard of chicken ranges from 40 to 150 mm Hg which is higher during the contraction of muscle laterales (thick muscle pair).

Factors affecting gastro duodenal motility

1. Contraction decreases during hunger and starvation.
2. Duration of contraction increases during hunger and starvation.
3. Duration of contraction increases on the consumption of fibrous and coarse feeds.
4. The amplitude of contraction of gizzard is increased by the presence of grits.

Nerve connection

The preduodenal parts comprising of oesophagus including crop, proventriculus and gizzard are innervated by the branches of parasympathetic nerve, vagus, the main motor nerve supply and also the sympathetic fibres. Motility is increased by the stimulation of the peripheral end of the vagus nerve and its ligation decreases.

Role of grits in gizzard

Hard grits are lodged in the gizzard. Grits are common in the gizzard of graminivorous, folivorous and herbivorous birds. It helps in the grinding of hard and coarse foods. However, it is not essential for normal digestion. Earlier it was a common practice to offer grits for free choice ingestion at the poultry farms which has been now omitted from the feeding system at poultry farms.

Deviation in gastro-duodenal motility of some birds

1. Duodenal contraction rate in turkey is 6-9 per minute of which initial 2-3 contractions are in rapid succession.
2. Mean amplitude of intra luminal pressure change in proventriculus of turkey is about 35 mm Hg.
3. The amplitudes of contraction of thick muscle of gizzard of foraging goose, common buzzard (wing soaring hawks) and great horned owl are 265-280, 8-26 and 60-175 mm Hg respectively.
4. The amplitude of contraction of thick muscle pair of the gizzard is significantly influenced by quantity of food eaten and gap between food intake and recording of pressure.

Post duodenal intestinal motility

There appears to be lack of information about post duodenal intestinal motility in chicken. Some sporadic studies mostly on turkey have provided limited informations, the important being the anti-peristaltic movement from cloaca and caeca for the absorption of moisture before exertion of urine.

The rate of contraction of ileum in turkey is an average 4 per minute. Mean amplitude is about 16 mm Hg. In some birds frequency of contractions may be upto 6 per minute of higher amplitude.

A peculiar motility of colon is the anti peristalsis which is probably a regular function. The functions are the transfer of urine from the cloaca into the colon and caeca for the absorption of water and second function is the filling of caeca. Anti peristaltic motility starts from the cloaca at the rate of 10-14 contractions per minute in fowl and turkey. Anti peristalsis stops just before defaecation which requires simultaneous contraction of entire colon.

Like colon two types of contractions also occur in the caeca of turkey. These are mentioned as minor and major contraction occurring at the rate of 2.6 and 1.2 per minute respectively. The minor contractions facilitate mixing of caecal contents and major contractions are peristaltic and anti peristaltic. Coordination of minor contractions occurs with ileal contractions and peristaltic motility in the colon. A series of major contractions, which are mainly anti peristaltic, are responsible for the removal of caecal contents and a single major contraction facilitates colon rectum evacuation or the process of defaecation.

The number of faecal droppings voided daily may be 25-50 including 1 or 2 caecal droppings. A diurnal rhythm may be observed in caecal motility being highest 1 contraction per minute during afternoon and lowest 1 contraction per 2 minutes during the resting period in night.

The duration of food passage through the alimentary tract depends on the composition of diets; it is longer in herbivorous birds and much shorter in carnivorous birds. Age may also affect the rate of passage. The food ingested at a time excreted at different times and may require about 24 hours in fowl. The excretion of orally administered marker may be detected in faeces in about 2.5 hours of feeding. However, almost all amount of marker except a very small quantity is excreted in 24 hours but traces of maker can be detected in caecal contents for 2-3 days. This indicates irregular evacuation of caecal contents in poultry.

Table 9.2: The anatomy and function of different segments of the digestive tract of the adult chicken.

Organs	Description	Function
Mouth	It is made of hard horny beak of upper and lower mandible and hard palate which communicates with nostrils through a slit. Teeth, lips and cheeks are absent tongue is narrow, soft palate is absent.	1. Food picking 2. Water intake, the head is raised for the ingestion of food, grit and water by gravity. 3. Rough posterior surface of tongue is useful for ingestion.
Salivary gland	Mulberry like irregular shape.	Lubrication for easy food passage to esophagus.
Pharynx	Narrow musculo-epithelial passage.	Communication for oral, digestive and respiratory organs.
Oesphagus	Long and flexible musculo-epithelial tube.	Passage of food and water to proventriculus partly direct and partly after stay in crop.
Crop	A diverticulum of lower oesphagus.	Storage and moistening of stored foods before passage to proventriculus.
Proventriculus (glandular stomach)	It is true or glandular stomach placed between esophagus and gizzard. Mucosal glands secret mucous and compound glands secret HCl, pepsinogen and mucous.	Secretion of gastric juice containing mucous Hydrochloric acid and pepsinogen. Gastric juice is mixed with ingesta. Pepsinogen is converted into pepsin. Digestion process starts.
Gizzard (Ventriculus)	Muscular stomach made of two pairs of muscles, the thin and thick muscle, may contain grits.	1. Physical grinding occurs. 2. Enzymic digestion rate increases.
Small intestine, Liver pancreas	First elongated U- shaped loop is duodenum. Remaining upto border of caeca is jejunum and ileum. Two bile duct drain into duodenum. Pancreatic duct open in distal part of duodenum.	The secretions are succus entericus from intestinal glands. Bile drains from liver pancreatic juice is poured by pancreatic duct.
	Numerous villi of small intestine provide large surface area for the absorption of nutrients liberated from foods in small intestine.	Two bile acids are taurocholic and glyeocholic that change from acidic to alkaline.Bile helps digestion and absorption of lipids and lipid soluble vitamins. Protein, carbohydrates and other food constituents are digested by the action of intestinal and pancreatic enzymes.

[Table Contd.

Contd. Table]

Organs	Description	Function
Caeca	Two blind pouches at the junction of small intestine and colon.	Microbial digestion of fiber fraction occurs. Volatile fatty acids produced are absorbed. Urine is pushed back by antiperistalsis for absorption of moisture part.
Colon	Collection of digesta.	Absorption of fluid fraction.
Cloaca	An annulated short terminal portion like rectum of higher animals. It is formed of coprodium, urodium and proctodium.	Collection of faeces and urine for void after feeling during the process of digestion.
Vent	External opening of cloaca controlled by sphincters,	Controls excretion

Excretion/defaecation

The greater part of fluid of digesta is absorbed through the pre-colon parts of the intestine followed by absorption through the wall of colon. The urine is flown to caeca by antiperistalsis for the absorption of fluid. The greater amount of digesta fluid is absorbed through the wall of the colon and the faeces is pushed back into the cloaca. This is quickly followed by the expulsion of apparently solid urine left after the absorption of moisture of urine through the wall of the caeca. The cloaca functions as rectum for the storage of faeces and urinary bladder for the storage of urine. The retention time of excreta (faeces+urine) in cloaca is very short and is excreted as droppings capped with a grayish white layer of urine.

The structure and functions of different segments of the digestive tract of chicken have been summarized in a tabular form (Table 9.2).

The secretions of digestive system and associated glands involved in the digestion of food in chicken

The various secretions actively involved in the digestion of foods are the saliva from the salivary gland, mucous from esophagus, gastric juice from proventriculus and succus entericus from the intestine. The associated digestive organs i.e., liver drains bile in the duodenum and pancreas drains pancreatic juice in the small intestine posterior to the opening of the bile ducts. The source, composition, site of action and various actions on the ingesta and digesta have been summarized in table 9.3.

Table 9.3: Summary of the source, secretion, composition and action of digestive secretions on the ingesta and digesta of the chicken.

Organ	Secretions	Constituents	Site of action	Main actions
Salivary glend	Saliva	Water, mucous, ptylin (amylase), salts.	Mouth, oesophagus including crop	Moistening and lubrication of food. Neutral medium for action of salivary amylase for hydrolysis of starch in to dextrin and maltose.
Crop (Lactobacillus dominate)	Bacterial enzymes	Salivary amylase bacterial enzymes mucous.	Crop	Action of salivary amylase continues. Bacterial enzymes produce lactic acid and acetic acid. Saliva mixed food is held for some time.
Proventriculus (i) Mucosal glands (ii) Compound glands	Gastric juice	Hydrochloric acid and pepsinogen	Proventriculus and gizzard.	Mixing of ingesta and gastric juice. Acid digestion pepsinogen becomes active pepsin.
Gizzard	–	Gastric juice, pepsin and ingesta. HCl	Gizzard	Physical grinding, mixing of HCl and enzymes, partial digestion.
Liver	Bile	Bile salts, bile pigments, cholesterol, water. Back flow of pancreatic lipase and amylase.	Duodenum	Neutralization of acidity of digesta, increase in alkalinity, emulsification of fat. Cleavage of fat into fatty acids.
Pancreas	Pancreatic juice	digesta, pancreatic amylase, lipase, trypsin, chymotrypsin	Small intestine	Digestion of starch, lipids and protein. Production of maltose and iso-maltose from starch and dextrin split of protein into peptone peptides and some amino acids.

[Table Contd.

Contd. Table]

Organ	Secretions	Constituents	Site of action	Main actions
Intestine (duodenal glands goblet cell	Succus entericus	water, mucous, entrokinase, carboxpeptidase, aminopeptidase dipeptidlase, maltase, isomaltase sucrase	Small intestine	Protection of enteric mucosa from acid damage. Split of polypeptides and dipeptides into amino acids. Splits maltose and iso maltose to glucose and fructose. Absorption of nutrients.
Caeca	Bacterial enzymes	Digesta, fibrous parts, fibrolytic enzymes of bacterial origin	Caeca	Digeston of cellulose and hemicellulose, production of votatile fatty acids. Absorption of moisture of urine received by anti peristalsis.
Colon	–	Residue and digested colon products.	Colon	Absorption of moisture, Volatile fatty acids.
Cloaca	mucous	Collection of faces and urine	Cloaca and vent	Excretion.

Digestion in chicken

1. Mouth is the organ of food and water intake.
2. Lips, teeth and cheeks are absent. Thus, there is no chewing of feeds.
3. Saliva containing amylase is mixed in the mouth during passage to oesophagus.
4. Taste glands are few and situated on the posterior part of the tongue and walls of the pharynx.
5. Oesophagus secretes mucous to facilitate passage of feed.
6. Part of ingesta is collected in the crop (in some species particularly grain eaters large amount is stored). Softening of feed occurs by saliva draining into crop and action of salivary amylase starts.
7. The feed directly from oesophagus and also from crop enters into the proventriculus or glandular stomach. The gastric juice containing hydrochloric acid and pepsinogen is mixed and then pushed back into the gizzard or ventriculus.

8. The ingesta is ground to fine chyme by the rhythmic contractions of thin and thick muscles producing high pressure ranging from 40 to 150 mm Hg which is higher during the contraction of thick muscle pair (musculi laterales)

   Koilin, a protein- polysaccharide complex substance is produced by the wall of gizzard. The amino acid composition of koilin is similar to keratin. Koilin hardens with the action with hydrochloric acid entering from the proventriculus.

9. Presence of grit is not essential in the gizzard but when present the efficiency of grinding is normally increased. Grits may increase the grinding rate of grains by 10 per cent approximately. It is more useful in foraging chicken.

10. Finely ground digesta along with the reflux of intestinal contents into the gizzard are pushed into the duodenum and small intestine.

11. Pancreas is situated in the loop formed by duodenum. The bile drains via the two bile ducts, one from gall bladder and another from the right lobe of the liver. Three pancreatic ducts open in duodenum posterior to openings of the bile ducts in the distal part of the duondenum.

12. Bile contains bile salts which are synthesized in hepatocytes from cholesterol. The second secretion is salt free fluid. The excretory constituent is bilirubin.

13. Small intestine is the main site of digestion. The media is made alkaline by bile and all necessary enzymes are secreted by the cells of intestine and also drained from the pancreas. Proteolytic activity is accomplished by aminopeptidases and carboxypeptidases secreted by the duodenal mucosa. Various intestinal enzymes secreted are amylase, maltase, sucrase and esterase. The pH of intestinal contents ranges from 6 to 8. Acidic metabolites of bacterial origin produced in the crop, caeca and colon lower the pH in these segments.

14. Ingesta entering in to the small intestine is digested by the enzymes of pancreas, intestine and microbial activities. The pH is maintained by the non enzymic portion of pancreatic secretion having buffering compounds. The protein is digested by trypsin and chymotrpsin. The other enzymes are dipeptidase, aminopeptidase and carboxypeptidase.

15. Pancreatic secretion rate is very high in chicken and many other birds when compared with mammals.

16. The bile drained into duodenum helps in neutralization of acidic chyme entering from the gizzard. Bile salts are used for the emulasification of fats required to facilitate fat digestion. Amylase activity has been also observed in many avian species. Almost 95 per cent bile salts are absorbed and reutilized for the synthesis of bile in the hepatocytes.

17. Digestion in caeca is little (about 10% total feed consumption) in the chicken and other members of galliformes. Caecotomy does not affect the life of chicken. Digestion of cellulosic fraction of the diet is carried by the cellulase and hemicellulase produced by microbial inhabitants of the caeca.

    Urine backflow is carried by antiperistalsis for movement from cloaca to colon and caeca for the absorption of moisture. It has been shown that moisture content of excreta increases after caecatomy. There is also evidence of vitamin synthesis in the caeca but probably it is not absorbed. However, scavenging birds may abtain by coprophagy.

Absorption of nutrients

1. Upper part of ileum is most active region for the absorption of the nutrients liberated during the digestion, specially the absorption of the end products of proteins (amino acids and small peptides), fat (emulsified fatty acids) and carbohydrates (glucose and fructose). Digestion end products of endogenous protein are absorbed through the walls of lower ileum.
2. Larger proportion of bile salts is absorbed through the walls of lower ileum.
3. Active absorption of D-glucose, D-galactose, D-xylose, alpha-methyl glucoside, 3- methyl glucose and also beta- fructose occurs.
4. Transportation of other simple sugars (monosaccharides) is generally passive.
5. Active transportation of monosaccharides in the small intestine of chicken is facilitated by the sodium dependent mobile carrier system. This system develops during the embryonic growth and completed before hatching.
6. Ability of glucose absorption is maximum in first week after hatching and there after decreases significantly
7. Amino acids transportation is also mediated by a carrier and rate is not affected by age. Proteinous substances are mostly transported in the form of small peptides from the intestine but appear as amino acids in the mesenteric blood.
8. Rate of transportation of neutral amino acids is higher than the acidic and basic amino acids.
9. There may be more than one system of amino acid transportation across the intestinal wall. Probably some of the amino acids are transported by more than one system. The transportation of glutamic acid may be either by carrier mediation or by diffusion but that of cystine is only carrier mediated route.

10. Antagonistic effect on amino acids absorption is seen in chicken. Presence of L-arginine, L-phenylalanine and L-histidine inhibits the transportation of L-lysine. Similarly L form of isollueine, methionine or valine inhibits the transport of L-leucine. In general D- amino acids inhibit the transport of L-amino acids through the intestinal wall.
11. The absorption rate of L-amino acids across the intestinal wall is not affected by molecular weight. The amino acids with long non-polar side chain are absorbed more efficiently. These amino acids are leucine, methionine and valine. The absorption rate of amino acids with polar side chains is slower.
12. The rate of digestion and absorption of the metabolites of carbohydrate and protein can be demonstrated by the estimation of the increased concentration of glucose and amino nitrogen in the blood of portal vein and wing vein in about 15 minutes of eating these substances.
13. The absorption of aldoses is active across the cells of microvilli membrane. The sugar is transported to liver. The ketoses (fructose) are also transported by a carrier but exact mechanism needs further exploration.
14. After digestion the soluble mixture of micelles is available for absorption. The transportation of micelles occurs by passive diffusion which is highest from the jejunum region or posterior part of the small intestine.
15. After absorption of micelles resynthesis of triacyl glycerol occurs. This requires energy (ATP) for reaction. The formation of chylomicrons of fat globules occurs. However, medium and short chain fatty acids do not require emulsification by bile salts or micelle formation and transported directly into the portal blood. The passage of these fatty acids is sodium dependent. It occurs against concentration gradient by active transportation into the portal venous blood in the form of low density molecules of lipoprotein. Role of lacteals in transport is very little.

Absorption of minerals

1. Absorption of liberated minerals occurs either by direct diffusion or by carrier mediated process. So far mechanism of transportation of all the minerals is not fully explored.
2. Absorption of calcium is regulated by vitamin D_3 (1, 25-dihydroxycholecal cipherol).
3. The 1, 25 dihydroxycholecalcipherol also enhances the absorption of phosphorus from the intestinal digesta. It also increases reabsorption rate of these two macro minerals from the kidneys.

4. Low pH increases rate of calcium absorption from the intestine.
5. Absorption of calcium from the intestine is inhibited by complex formation with oxalic acid and phytic acids by forming insoluble oxalates and phytates. Phytic acid inhibits absorption of several minerals.
6. Absorption of calcium is also influenced by physiological state of the body. In case of laying birds absorption of calcium is very high due to active function of the shell forming glands.
7. Iron absorption is mostly independent and rate is regulated by iron content in the body because excretion is negligible. However, absorption increases in anemia.
8. Absorption of zinc from small intestine is a carrier – mediated system.
9. Inorganic iodine is more efficiently transported when compared with the organic compound of iodine.

Absorption of Vitamins

1. Fat soluble vitamins are transported with fatty acids diffusion across the intestinal wall into portal blood, and then carried to liver.
2. Absorption rate of vitamin A is much higher than the absorption of carotenes. However, diets containing more than 30 per cent yellow maize has been found to meet the vitamin A requirement of chicken.
3. Conversion of ester form of vitamin A to alcoholic form (retinol) by irradiation is required for efficient transportation.
4. Phytoesterols require irradiation for conversion to vitamin D for efficient absorption.
5. Water soluble vitamins are transported by diffusion and carrier-mediated process.
6. A glycoprotein is responsible for the transport of vitamin B_{12}. the carrier is called intrinsic factor.
7. Vitamin K is synthesized by the intestinal bacteria.

Summary of digestion and absorption

The summary of digestion of foods, production of absorbable end products and site of absorption are presented in table 9.4.

Table 9.4: Summary of digestion in poultry.

Nutrient in food	Products of digestion	Site of absorption
Carbohydrates	Glucose	Intestine(mainly ileum)
Fibrous Carbohydrates	VFAs	Intestines (caeca & colon)
Protein	Amino acids & small peptides	Small intestine
Lipoids	Monoglycerides	Small intestine
Minerals	As elements	Small intestine
Vitamins	Vitamins	Small intestine

Metabolism of nutrients

After digestion of feeds in the alimentary canal the nutrients liberated are absorbed in the fluid compartments of body, mainly blood circulation. The absorbed nutrients aborbed are transported to specific organs for metabolism, like liver, muscles, long bones and spleen for the maintenance of vital functions, storage of surplus and elimination of unwanted metabolites in faeces, urine, expired gases etc.. The nutrients available from the digestion of dietary carbohydrates, lipids, proteins, minerals and vitamins are summarized in Table 9.5.

Table 9.5: Forms of nutrients absorbed for metabolism in the body and fate of metabolism.

Nutrients in feeds	Form of nutrients for metabolism	Fate of metabolism products
Carbohydrates	Simple sugars, main is glucose.	Energy liberation for body functions, formation of ribose for DNA and RNA, glycogen, fats and carbohydrates of mucous and cerebrosides.
Lipids	Fatty acids, glycerol and fat soluble vitamins	Supply of essential fatty acids, maintenance and formation of cell membranes, gluconeogenesis for energy supply, formation of steroids etc.
Proteins	Amino acids	Formation of proteinous structural tissues of body like muscle, bones, epithelium, dermis, fibres, glands, feathers, gonads etc. Synthesis, of hormones, enzymes, blood etc. Gluconeogenesis in non-availability of primary sources.
Minerals	Ionic form and organic salts.	Skeletal growth, repair of wear and tear. Acid base balance, maintenance of fluid tonicity, muscle contraction, transmission of nerve impulse etc.
Vitamins and precursors	Vitamins and precursors	Cofactors in enzyme systems. Maintenance of integrity of membranes. Maintenance of growth, production, reproduction and health.

Energy metabolism

The carbohydrates transported in to the body system are the primary sources of metabolizable energy (ME) necessary for the body function. Glucose is the main metabolizable sugar always circulating in the blood. A fall in blood glucose level below the normal value (130-270 mg/dl) the mobilization of body fat followed by body protein begins to maintain normal blood glucose level by gluconeogenesis.

The energy yielding nutrients (carbohydrates, fats and proteins) are metabolized through the common tri carboxylic acid (TCA) cycle. The calorie availability as ME is 3.33 and 3.25 kcal per gram of glucose and fructose respectively. Metabolism of fat yields about 8 kcal ME per gram. Energy supply by protein (amino acid) is uneconomical resource utilization for body functions.

Calorie: protein ratio

Calorie means nitrogen corrected apparent metabolizable energy (AMEn) and protein means the percentage of protein in the diet where protein is balanced for the essential amino acids. Therefore for efficient utilization of dietary energy and protein in the body it is necessary to maintain the ratio of AMEn and balanced protein. The values are different for parent stock, growing broilers, finishing broilers, pullets and layers of different avian species.

Distribution of ingested energy in the body

1. Maintenance of blood sugar.
2. Carbohydrates of cells.
3. Storage in the form of glycogen in liver and muscles.
4. Carbohydrates of eggs and sperms.
5. Body fat and egg fat production.
6. Oxidation in body for maintenance of body functions and normal body temperature.

Fat metabolism

Fat globules and fatty acids in the body of poultry are metabolized for the following functions:

1. Oxidized in TCA cycle for the release of heat necessary for body function.
2. Deposition as subcutaneous adipose tissue.
3. Deposition as body fat.

4. Transported to functional ovary (normally left ovary) for the formation of yolk fat.
5. Transportation of fat soluble vitamins in the body for function at target organs.
6. Supply of essential fatty acid (linoleic acid) which affects the size of egg yolk and whole egg in chicken.
7. Gluconeogenesis for body functions during deficient supply of carbohydrates.

Protein metabolism

Optimum supply of essential amino acids is necessary for proper growth and egg production. The normal functions of amino acids are:

1. Formation of tissues.
2. Repair of tissues.
3. Formation of eggs.
4. Deamination and deposition as fat in depots.
5. Gluconeogenesis in deficient energy producing nutrients (carbohydrates and fat).

Essential VS non-essential amino acids

Body tissue synthesis requires both essential and non- essential amino acids at the optimum calorie: protein ratio. A short supply of non- essential amino acids in low protein diet will cause wasteful utilization of essential amino acids (EAA) for the synthesis of non-essential amino acids (NEAA) necessary for the maintenance of EAA: NEAA ratio required for body tissue formation. Feeding of a diet of suboptimal protein and higher EAA:NEAA ratio (65:35) reduces feed intake causing reduced ME availability and lowered production due to loss of energy in deamination and utilization of energy for the synthesis of non- essential amino acid (s) in deficit supply. Optimum EAA: NEAA ratio at optimum protein: energy ratio ideal in standard diets should be about 55:45 for growing bird and 50:50 for laying hens.

Proline is considered a non- essential amino acid for the poultry is synthesized in smaller amount than the requirement of high egg laying birds. Therefore it is a conditional essential amino acid and it may be more beneficial to supply proline in feeds of laying hens.

10 POULTRY FEED PROCESSING TECHNOLOGY

Any naturally available feed ingredient is not balanced to meet the optimum nutritional requirements of poultry and other avian and mammalian species. Therefore, several feed ingredients require mixing for the supply of required amount of nutrients for various physiological functions. Since feed ingredients are available in different form varying in volume and density it is difficult to maintain homogenesity of compounded feeds. Therefore, feed ingredients are processed for assuring uniform mixing and homogenesity of ingredients in the finished products ready for feeding and marketing.

Personnels associated with feed processing

Feed processing is a coordinated work of selection of feeds, milling of feed ingredients, quality assessment of feed ingredients before milling, during milling and after milling and calculation of the ratio as well as actual quantity of various feed ingredients for mixing and further processing. For these works services and expertise of following personnels are required.

1. Poultry nutritionist/Animal Nutritionist
2. Instrumentation engineer with specializing in operation and maintenance of feed processing plants.
3. Feed processing technologists
4. Skilled man power for the running and maintenance of feed mill.
5. Unskilled workers
6. Feed mill manager

Points to be considered before decision making for the establishment of a feed mill

Compounded feed production is an exclusive commercial venture and before establishment of a feed mill for the production of compounded feeds it is important to conduct SWOT analysis. This will provide necessary informations regarding strength, weakness, opportunity and threat in the business in the area of operation.

1. Area for the business.
2. Market capacity of the area and scope of further development. This is required for determining the size of plant.
3. Types and size of poultry farms in the area and growth pattern of poultry in the area during the past 3-5 years.
4. Growth pattern of poultry products (eggs and meat) consumers in the area during the preceeding 3-5 years.
5. Capacity, function and status of the other feed mills in the area.
6. Sources of compounded poultry feed supply in the area.
7. Production of basic food ingredients like maize, other cereal grains, cereal milling by-products (Wheat bran, deoiled rice bran) and oilseed meals in the area.
8. Transportation facilities for bringing essential feed supplements and additives not manufactured or produced in the area like fish meal, mineral mixtures, vitamin supplements, yeast and other feed additives.
9. Availability of skilled and unskilled manpower for providing paid services in the feed mill.
10. Basic infrastructure facilities like rail, road, electric supply and water supply.
11. Land availability for establishment of feed processing plant.
12. Socio-political interference in the working of other mills in the area.
13. Financing institutions (Bank) in the area.

On the basis of these informations scope of running a poultry feed mill can be decided and economic feasibility can be worked out with the help of marketing and accounts personnels.

Roles of specialists and other personnels of the feed mill

Although theoretically roles of different technical personnels may be considered independent but in actual functioning coordination is the key of success.

Role of mill manager

Mill manager has pivotal role around which the entire machinery rotates. Some of the important roles of feed mill manager may be listed as follows:

1. Maintenance of office, records and warehouses.
2. Procurement of ingredients and supply of feeds against indents.
3. To meet the material requirements of entire work force
4. To keep an eye on the financial status of mill through frequent/periodic review of the market.
5. Maintenance of harmonious working condition in the mill.
6. To provide satisfactory services to consumers (poultry farmers and dealers) regarding punctuality in assured supply and maintenance of quality.
7. Proposing incentives to personnels for good performance higher, output and more profit.
8. To take care of welfare activities for children education, health, recreation and community get together etc.

Role of poultry nutritionist

The main role of a poultry nutritionist is to assure supply of wholesome compounded feeds for different categories of birds. This involves following works:

1. Procurement of wholesome feed ingredients on the basis of physical and chemical analyses.
2. Procurement of feed supplements and feed additives from standard manufacturers.
3. Utilization of perishable ingredients before expiry date.
4. Formulation of different categories of compounded feeds from the available ingredients. Use of many ingredients may produce following problems:
   (a) Error in weighing in different batches may alter composition.
   (b) Difference in particle size due to hardness of different grains/ feeds may disturb homogenous mixing causing variation in nutritive value of samples from different strata of the lot.
   (c) Mixing time increases.

   (d) More space requirement for the storage of many feed ingredients.

   (e) Problems in homogenisity and conditioning of pelleted feeds.

5. Special attention is required during procurement, handling and milling of dried whole fish because dried whole fish available in the Bhartiya markets is prepared by Sun drying on the hot sand near the sea sore. Large quantity of common salt is also sprinkled during drying for enhancing the process and reducing the loss due to autolysis. Thus, milling of such dried whole fish increases wear and tear of grinder. In many feed mills small stone grinders are used for grinding of fish.

6. Quality control at all stages right from procurement of feed ingredients to production of finished product is the exclusive responsibility of poultry nutritionist of the feed mill.

Role of mill maintenance Engineer

The main role of a feed mill engineer is at the time of installation and commissioning of the feed milling plant for the following purposes:

1. Determination of the size of plant on the basis of growth pattern of poultry farming in the area and scope of feed supply in the other areas.

2. Selection of feed milling plant and determination of level of automation for reducing dependence on work force.

3. Selection of suitable sites for the installation of milling plant, godowns for feed ingredients, godwons for finished products, workshop for the maintenance of plants within the boundary of the mill campus.

4. The engineer should be reasonably conversant with different disciplines of engineering required for the smooth functioning of feed mill.

   (a) Handling of electrical parts like motors, switchgear equipment and wiring etc.

   (b) Operation and maintenance of boiler and steam supply alongwith associated activities like maintenance of steam traps and pressure controlling valves etc.

   (c) Operation and maintenance of air compressor and pneumatic cylinders, and also the maintenance and replacement of different parts like filter, regulator and lubricator etc.

   (d) Maintenance and replacement of electronic equipments like weighing machines and sac stitching machines.

(e) Maintenance of safety during the operation. This needs special attention and daily thorough checking specially of sensitive points before the start of operation. In shift working there should be thorough inspection at the end of previous shift and start of next shift. Rest period for the machine is decided by the engineer.

Role of feed processing technologist

This person is responsible for running the feed mill in consultation with the engineer. He is always present during the working period. His jobs may be counted as:

1. Operation of feed mill.
2. Observations of functioning of different parts during working.
3. To assist the engineer in selection of parts for replacement or modifications/alterations for changing the production capacity of the plant.
4. To decide milling and mixing of different feed ingredients and conditioning of pelleted feeds in consultation with the poultry nutritionist.
5. Proper utilization of electricity, water, boiler and other items responsible for adding insensible cost in the production cost.
6. Maintenance of cohesive relationship among work force.
7. Maintenance of sanitary condition.
8. Maintenance of a functional mini workshop for handling routine works of minor nature.
9. Control of insects and pests particularly rodents, termites, tribolium and birds in the plants and godwons.
10. Preparation of production plan in consultation with mill manager, poultry nutritionist and engineer.
11. Maintenance of inventory.

Role of skilled workers

These are trained and/or experienced persons in different trades necessary for the smooth running of a feed mill. Some skilled workers may be capable of handling more than one operation in the mill. These persons are responsible for the smooth functioning of work like grinding, mixing, steam supply, pelleting and conditioning of pellets etc.

Role of unskilled workers

This group is responsible for working as per the instructions of the superiors. However, some routine works are done daily and do not need specific instructions. These are weighing, loading, unloading, shifting, cooling of pellets, collection of pellets, filling of sacs, weighing and stiching etc. However, marking is done by skilled workers and verified by the superiors.

Requirements of establishing feed mill

1. A suitable land at high level free from the menace of water logging and flood. It should be at least 1 km away from the human habitation and 2-3 km away from the hospitals and educational institutions.
2. The place should be connected with road and have accessable electric supply and water supply.
3. Area should be properly fenced and have enough space for the construction of office, warehouses, weigh bridge.
4. Space for feed mill and associated accessories like boiler, power generation equipment and mini workshop equipped with essential appliances and materials.
5. Feed analysis laboratory equipped with essential instruments, glasswares, plasticwares and chemicals etc.
6. Office and office store
7. Retiring/changing room with separate toilets for ladies and gents for out door workers.
8. Housing facility for residential employees.
9. Transportation facility for weekly marketing and emergency.
10. Plantation for minimizing environmental pollution.

Requirements for the functioning of feed mill

The requirements for the operation of feed mill may be listed as follows:

A. Man power

1. Feed mill manager
2. Plant engineer
3. Poultry nutritionist

4. Operational technician
5. Laboratory assistant
6. Plant running technician
7. Skilled and unskilled manpower

B. Essential supplies and facilities

1. Feed ingredients
2. Assured power supply
3. Running water supply
4. Transportation facilities for quick movements.
5. Maintenance equipments in working order
6. Good quality first aid kit and a trained health assistant.
7. Canteen as per the strength of staff.
8. Adequate security staff from a registered security providing agency or direct recruitment from retired staff of forces.

Types of poultry feeds

For the feeding of poultry compounded feeds are prepared either in mash form or pelleted feeds.

1. Poultry mash

Mash feed is a balanced dry mixture of essential feed ingredients prepared as far as possible by uniform mixing of coarse ground feed ingredients. Now a days poultry mash feed is prepared at small scale by the local manufacturers catering the small requirements of small scale poultry farmers keeping few birds in backyard for domestic uses or 50 to few hundred birds for subsidiary income.

Method of preparation

The feed ingredients are procured and coarsely ground. The quantity of each ingredient is weighed and piled in a sequence of largest quantity as a basal layer followed by amounts of other ingredients in descending order. Supplements and additive constituting less than 1% are mixed with some amount of other ingredients used in larger quantity and placed as a top layer. By this manual mixing method only 1-5 quintal feed mixture is prepared at a time. The pile of feeds is mixed with the help of spade or belcha first in order of column and thereafter 5-6 turnings of entire lot is given to ensure uniform mixing. After mixing mash is stored in bins or sacs at a dry place.

Large batches of more than 25 quintals are mixed mechanically in feed mixer, which are fabricated by local artizons in almost all states or purchased from northern part.

Advantages of mash

1. Always fresh feed ingredients are used
2. Easy to prepare.
3. Require few low cost equipments.
4. Small space is required for preparation and storage of feed mixture.
5. Poultry farmers can prepare after obtaining feed formulae from a poultry nutritionist of government institution.
6. Financial requirement is small and payment of bank loan is easy and interest load is small due to quick turn over.

Disadvantages of mash feeding

1. Wastage by birds is more.
2. Distribution of micro nutrients does not remain uniform because higher density minerals gradually settle down. Therefore, mixing before feeding is required.
3. Fluctuation of feed quality is more due to frequent purchase of ingredients.
4. Some times either some ingredient is not available in the market or price has risen enormously.

2. Pellet form of poultry feed

Pellet making of compounded feed is done by special machines fabricated for the purpose and operated by electric power. For the feeding of poultry pellets of 3 and 5 mm diameter are prepared by a feed pelleting machine. Sequence of operation is in following order:

(i) Procurement of feed ingredients

(ii) Weighing of required quantity of feed ingredients.

(iii) Grinding of feed ingredients

(iv) Mixing of feed ingredients as per the formula provided by the poultry nutritionist

(v) Conditioning of feed mixture with steam

(vi) Pelleting

(vii) Cooling of pellets

(viii) Coarse crumbling

(ix) Packaging

(x) Inscribing details of feeds on package.

(xi) Storage

(xii) Shipment/marketing

Some other methods of compounded feed production

In an attempt to reduce the cost of production of poultry feeds in an environment of continuous change in the cost of not only the feed ingredients but also the other essential inputs like electricity, petrol, deseil, water supply, transportation and work force the technical personnels are continuously working for the utilization of greater proportion of various non-conventional feeds. In this endeavour some other methods of preparing compounded poultry feeds have been developed. Some of these are compaction, double pelleting, expansion and extrusion.

1. **Compaction:** It is a process of densification of mixed feed for greater fusion for reducing the shorting of any feed ingredient (if less palatable). The compaction is a process of application of pressure for the reduction of particle size, heat treatment and moisture or steam application (if necessary). A compactor is used for the purpose which works in combination with the pelleting machine. Compaction of feed mixture precedes pelleting.

Advantages of compaction

1. Improvement of feed quality.
2. Increased use of relatively less palatable non-conventional feed ingredients.
3. More hygienic compounded feed production.

4. Improvement of palatability
5. The mixture can be fed as compact feed or pelleted, feed.

Double Pelleting

In this process pellet is passed again through the pelleting machine for heat treatment produced by friction.

Advantages

1. Volume is reduced significantly.
2. Increased density often increases voluntary intake.
3. Storage space and handling processes are lowered.
4. Per unit space greater amount is transported.

3. Expansion

It is a combination of compaction and pelleting for improving the palatability and feeding value of the product. Output of machine is increased.

4. Extrusion

This is a process of expansion of surface of feeds for providing larger exposure for the action of digestive enzymes and increased availability of the nutrients. The extrusion is completed in the sequence of grinding, mixing, moistuning, cooling, expansion, dehydration or drying and stabilization.

Processing procedure of extrusion: In this process feed mixture is subjected to high friction and shearing in a strong metallic barrel of short length. For this purpose a series of screws and frictions placed at short distance for action and counteraction. The feed mixer is pushed forward by the screws while forward flow is inhibited by the restriction of the frictions. These processes produce high temperature and pressure inside the barrel. This is followed by the release of extruded feed through a nozzle on a clean and dry surface or heavy duty polyethylene sheet for cooling and packaging.

Since the barrel is short, inside temperature rises to about 180°C for a short period of 30 seconds. This is like high temperature short time (HTST) method of pasteurization. Heating at such a high temperature only for 30 seconds helps satisfactory sterilization of feed without sensible damaging effect on nutritional quality.

Establishment of poultry feed manufacturing mill

Once the decision is taken on the basis of SWOT analysis, the process of acquiring land, construction, procurement of machinery and appointment of manager, engineer and processing technologist is completed.

Main components of a feed mill

For the construction of commercial feed mills the main components are almost common and difference is only in the size according to production capacity. The main components of a feed mill are:

1. Grinder, 2. Mixer, 3. Pellet mill, 4. Pellet cooler, 5. Pellet crumbler and 6. Sifter

1. Grinder

Feed grinding is the process of reducing particle size as per the need of the feed to be manufactured. The preferred equipment for feed grinding is a hammer mill with swinging hammers. Screens of different size are used for obtaining desired particle size.

Points to be considered for the selection of grinder

1. Maintenance of particle size as per specifications.
2. Power utilization efficiency
3. Easyness of changing replaceable parts like screen, hammer and other parts which wear frequently.
4. Speed of hammers in terms of meters per minute.
5. Vibration and intensity of noise produced.
6. Average functional life of wearable parts and assurance of supply or alternate sources in the market.

2. Mixer

Mixer is an important component of feed mill and homogenous mixing of feed ingredients is essential for the maintenance of quality of the final product, i.e. the compounded feeds in mash or pellet form. Different types of mixers used in feed mills are the vertical types and the horizontal types. The later type, viz. horizontal paddle type, horizontal double shaft paddle type and horizontal ribbon type are preferred by the feed mills.

Points to be considered in the selection of mixer

1. Power utilization efficiency.
2. Time taken for mixing of each batch.
3. Uniformity of mixing of feed ingredients.
4. Clean out efficiency, i.e. residue in mixer should be minimum.
5. Average life of parts requiring frequent replacement due to wear and tear.

At present most efficient mixer is double shaft paddle type followed by horizontal ribbon type, horizontal paddle type and vertical mixer in descending order as evident from the relative performance compared in Table 10.1.

Table 10.1: Comparison of mixing efficiency of different mixers.

Types of mixer	Batch mixing time (Minutes)	Coeffient of variance
Vertical mixer	20(15-30)	>20
Horizontal mixers	1-10	
(i) Paddle	8	13 (12-15)
(ii) Ribbon type	3	9 (8-10)
(iii) Double shaft paddle	1	4 (2-5)

3. Pellet making machine with conditioner

The rate and uniform distribution of steam supply from the boiler affect the conditioning of mixed feed transferred from the mixer. After conditioning mixed feed is compressed with the application of desired pressure and passed through the die of predecided size for pelleting. The pellets delivered through the die are simultaneously spread for drying or passed through a drier for efficient and uniform drying.

Points to be considered for selection of pellet mill

1. Power consumption per unit of pellet produced.
2. Design and efficiency of conditioner. This is important for uniform injection of steam required for proper moistening to facilitate compactness of pellets.
3. Longivity of die and rollers.
4. Uniformity of hole size and hardness of die.
5. Durability of pellet die.
6. Pressure bearing capacity of pellet die.

7. Easy die replacement in the machine.
8. Easy lubrication of mobile/ rotary parts.
9. Minimum left over unpelleted mash.
10. Stability of pellets.
11. Availability of accessories requiring frequent replacement due to normal wearing.

4. Pellet cooler

The device used for quick cooling of pellets for avoiding spoilage due to microbial and mould growth. Three types of pellet coolers used are counter flow type, horizontal type and vertical column type. During the process of conveying forced cool (normal) air is passed through the pellets for the evaporation of moisture in the pellets. This also produce cooling effect due to loss of heat in evaporation of pellet moisture. The efficiency of cooler is determined on the basis of cooling effect which should be across the pellet upto core otherwise moisture may remain in the core region. Incomplete cooling causes sweating of the pellets means passage of moisture from core to surface. This may cause lump formation, caking and mould growth. Therefore, slow cooling is required.

Points to be considered in the selection of pellet cooler

The cooling efficiency of pellet cooler depends on the following conditions:

1. Residence time: It is time taken by pellets to stay in cooling chamber for complete cooling or brining down the pellets at room temperature.
2. Cooling efficiency of machine declared by the manufacturer.
3. Air used for aeration of pellets for cooling. Dry air should be circulated for fast cooling. Humidity in air passed through pellets will decrease the process of cooling and high humidity in passing air may not be able to do the job of moisture evaporation from the pellets. Occurrence of clumping due to slow cooling may increase which is not desired.
4. Design of discharge chamber and cycle determines efficiency of cooling.
5. Materials used for fabrication, sturdyness and service life of pellet cooler should be of good quality.
6. Functional efficiency of automation for the discharge of cooled pellets.

   Generally counter flow pellet cooler is used in the pellet feed manufacturing mills.

5. Pellet crumbler

The machine used for breaking the pellets into small pieces is known as pellet crumbler. The size of crumbling is decided on the type of birds to be fed. Small pieces are required for the feeding of young chicks in early life. Older chicks and adult birds can eat larger pieces or can break with the help of beak and claw for feeding. Pellet crumbling is done by rollers moving in opposite direction and gap between the moving rollers is fixed by device provided in the crumbler for crumbling pellets to desired size. Pellets to smaller diameter can be prepared but it has been found uneconomical. Feed mills generally produce pellets of 4 mm diameter.

Points to be considered for the selection of pellet crumbler

1. Materials used for the fabrication of rolls, rate of wear and duration of average service, i.e. quantity of pellets crumbled during working life.
2. Consistency in the size of broken pellet particles. Dust production should be minimum.
3. Easyness of cleaning and
4. Simple device for the adjustment of gap between the moving rolls.

6. Sifter

The machine used for the grading of pellets on the basis of size after crumbing is called sifter. The pellets or crumbled pellets can be separated into two or more grades of size with the use of sieves/screens of different size. Different types of sifters used in feed mills are reciprocating and vibrating etc. The portion of pellets or crumbled pellets passing out through the sieve is discarded for marketing and generally recycled with feed mixture for pelleting.

Points to be considered in the selection of sifter

1. Durability of the machine
2. Service quality
3. Easyness of changing sieves
4. Vibration adjustment device
5. Overall working efficiency

Material handling equipments and appliances

During the process of feed mixing almost all operations of handling are done manually in very small local organizations. The equipments used are simple flour mills with adjustment for cracking and crushing cereal grains and oil cake flakes. The other appliances are belcha, spade, buckets, gunny bags, weighing beam balance or small electronic balance of half or one tone capacity.

Equipments required in medium and big feed mills

Material handling equipments used in feed mills are:

1. Bucket elevators, 2. Screw conveyors, 3. Chain conveyors, 4. Bins, 5. Batch weighing and processing control machine, 6. Bag weighing machine, 7. Minor equipments like belcha, spade and buckets etc. 7. Tool box (small) and 8. Wheel barrow or trollies etc.

Boiler for steam generation with device for steam injection

Steam is required for the conditioning of mixed feed before passing through pellet making machine. Various types of boilers using different sources of energy may be used. The common sources of energy are electricity, diesel, gas, coal, rice husk and sugar cane bagasse etc. Now a days restrictions imposed on the use of burning organic fuels, use of electricity and gases is increasing.

Generally coil type, shell type and tube type boilers are used. Usually steam at 2-3 bar pressure is used for the conditioning of feed mixture before pelleting. There should be device for the adjustment of dryness of the steam. For good quality steam production water used should normally possess the quality shown in Table 10.2.

Table 10.2: Optimum water quality for good steam generation in boiler.

Sl. No.	Quality attributes	Values
1.	pH	8.5 – 9.5
2.	Dissolved solids	400 mg per litre
3.	Turbidity (NTU)	1mg
4.	Alkalinity	<250 mg
5.	Carbonates of calcium	<30 ppm
6.	Sulphates as calcium salts	<300 ppm
7.	Silica as oxide	<25 ppm
8.	Iron	<0.1 ppm
9.	Total hardness measured as calcium carbonates	<5 mg /litre

Accessories of feed mill

1. **Magnetic separators:** It is not uncommon to find iron wire and other iron pieces of different shape and size alongwith other impurities in the sacs of feed ingredients procured. Impurities of iron pieces are more due to extensive use of equipments of iron in agricultural operations. These pieces are damaging for the machine and birds eating the feeds. Thus, for actual location and efficient removal of iron pieces and particles magnetic separators are ideal. These are used for iron pieces separation from the lots of ingredients before grinding and again before placing the feed batch for pelleting. For maintaining the efficiency the magnets of separator should be cleaned off isolated iron materials from the feeds.

2. **Fluid coupling:** Such couplings are generally used in batch mixer when operation of mixer on full load is required. This is required in power failure during working of the mixer for mixing a batch.

3. **Variable speed drive:** This device is required only in feed mills used regularly at different speeds of passage of batch through the grinder and pellet making machines.

4. **Dust extraction device:** For smooth working, minimizing health hazard and environmental pollution due to emission of dust particularly during grinding and crumbling, dust extraction is required during the operation. Therefore, installation of dust extraction devices is required at such points in the feed mill. The machine sucks dust particles from the operation area and collect in a filter bag for disposal at a suitable place. Dust extraction machine should be easy for use and cleaning for smooth working.

Storage facility

Godowns and warehouses are required for the storage of feed ingredients and the finished products. Since many important ingredients like cereal grains and oilseed cakes are seasonal products their availability at competitive rate is more and assured. In order to maintain the quality of final products now a days seasonal procurement of feed ingredients specially yellow maize, white maize and other cereal grains is increasing. Thus, requirement of storage facility is also increasing.

Two types of storage facility are required in a feed mill for (i) storage of feed ingredients and (ii) storage of the final products for marketing. Separate warehouses for cereal grains, oilseed cakes and minor constituents like mineral mixture, vitamin supplements and feed additives are required. The capacity or size of different ware houses is determined on the basis of space requirement and frequency of procurement of the basic feed ingredients during a year or so.

Infrastructures for storage

1. Damp free silo construction at a high level for avoiding shipage and water logging.
2. Godowns of corrugated iron sheets, asbestos sheets or cement-concrete etc. for the storage of gunny bags or sacs filled with feed ingredients.
3. Godowns for storage of finished products. The quantity of finished products and duration of storage is much shorter.

Some times temporary storage for short duration may become necessary due to availability of raw materials at competitive price or short duration delay in marketing. For such situations a reasonably raised pucca platform and adequate terpoline should be available at the mill or can be hired at short notice.

Miscellaneous equipments and appliances

The other important equipments and appliances of a feed mill are:

1. **Electrical switch gear equipment:** There should be clear and well defined map of electrical wire fitting with depiction of important points and connections. Some of the important considerations are the use of switch fuse units, circuit breaker, motor starter and use of Direct –on-line (DOL) only upto 7.5 HP motor rating, star-delta used in motors of 10 HP and more upto 150 HP. This is a two stage starter initiated by "star" and then switched to "delta". For switching over from "star" to "delta" three contactors are placed along with a delay timer required for controlling time gap necessary for switching from "star" to "delta" starter.

   Soft is used for motor power rating 100 HP and more, but may be used for 50 HP. This system of starting reduces current consumption which becomes one-third to half.

2. **Hyrdaulics and Pneumatics:** These are equipments required for movement with the application of pressure.

Maintenance of Feed Mill, Accessories and all associated structures of the factory

This is most important aspect of running a feed manufacturing plant at a reasonable and sustainable profit. Maintenance of a feed factory should start from the boundary wall and finish at the feed delivery points. It means.

1. Maintenances of premises.
2. Maintenance of infrastructure
3. Maintenance of different components of feed mill and accessories.
4. Cleaning and disinfection of warehouses and godowns etc.
5. Manuals of each equipment and appliance should be kept intact at a proper place for reference and consultation.
6. No laxicity can be allowed in maintenance of electrical fittings and operation of machines that may be dangerous for life during failure.

A small workshop with a group of trained persons in handling electricity, electronic equipments and components of different machines and accessories should be either employed or should be available in short time on telephonic call. Annual contracts for maintenance of important equipments requiring advanced engineering services may be made with the manufacturing firms or dealers etc.

Quality control of poultry feeds

The success of feed production industry depends on the maintenance of high quality of finished feeds. Judgement of feed quality by most of the poultry farmers are the productivity, feed conversion efficiency and ultimately return on the expenditure.

Quality of compounded feeds depends on the quality of feed ingredients selected and processing technology for producing the final feed as per the nutritional parameters provided by the poultry nutritionist. Two main criteria are used for the assessment of quality.

1. Physical appearance and
2. Nutritional composition

Important quality control measures may be listed as follows:

1. Inspection of feed ingredients for soundness, uniformity of shape, size and colour, presence of adultrants and undesired harmful factors like rodants droppings, insects boring like tribolium infestation, droppings of stray birds.
2. Purchase of feed ingredients as per density and physical quality.
3. Collection of representative samples for chemical analysis.
4. Chemical analysis of each ingredient and final products before marketing. Special attention is given for fibre, fat, acid insoluble ash in routine analysis.

In case of use of non-conventional feeds and partially damaged organic feed ingredients detail analysis of amino acids particularly the limiting amino acids like lysine, methionine, threonine and tryptophane may be required.

5. Detection of mycotoxins and other harmful factors.
6. Maintenance of hygiene and sanitation in the premises, mill and appliances etc.
7. Proper storage in dry conditions in godowns free from insects, rodents and stray birds etc.

Packaging and labelling

The final products or compounded feeds in the form of mash or pellets are packed in containers of 10, 50 and 100 kg or as per the norms of a factory for marketing.

Materials used for packaging: Sacs/bags of jute or synthetic fibres of different capacity are used for the delivery of compounded poultry feeds.

Labelling of compounded feeds: Proper labelling is done on the body of sacs or bags. Label depicts the following or similar informations.

1. Brand name of the feed
2. Type of feed – for chicks, grower, layer or broiler
3. Quantity in kg or lb on the date of packing.
4. Date of manufacture depicting date, mouth and year.
5. Date of expiry, if any, depicting date, month and year.
6. Instructions for storage and feeding.
7. Chemical composition on dry matter basis, viz.

   Maximum moisture %

   Minimum protein %

   Minimum ether extract %

   Maximum fibre %

   Maximum acid insoluble ash %

   Informations about trace minerals and vitamins.

11

FEED FORMULATION FOR POULTRY

Balanced feeding is the key of efficient poultry production and compounding of cost effective palatable balanced feeds is the key of remunerative poultry production. Feeding cost of poultry production is 70-75 per cent of total production cost. Competition for cereal grains is increasing between humans and poultry in this country due to increasing human population and increasing demand of poultry products and more or less stagnation in the production of cereal grains, specially the yellow maize around which poultry production grows. The production of proteinous feeds for poultry (soyabean meal, groundnut cake, cotton seed cake etc.) may meet the requirements of poultry industry but often stress is created by export of protenious feeds.

Feed formulation for poultry is an art of application of nutritional knowledge for the compounding of balanced feeds. The use of mathematics helps in finding out the exact quantity of different feed ingredients of the desired formula.

Points to be considered for feed formulations

1. Availability of common feed ingredients.
2. Assured supply of feed supplements and feed additives
3. Economic consideration in the selection of feed ingredients without any adverse effect on the health, production and quality of poultry products. For example, feeding value of poultry does not change on replacement of yellow maize with white maize but often value of white yolk eggs is less due to lack of pigmentation of yolk although adequate vitamin A is supplemented.

Balanced compounded feeds

Balanced compounded feeds of poultry are generally supplied in the form of mash, pellets and extruded feeds. Marketing of mash has significantly decreased and now mash feeds are limited to small scale local feed manufactures and small poultry farmers. Now a days pelleted poultry feeds are more popular due to easiness of feeding and little wastage during feeding.

Components of poultry feeds

The main components of poultry feeds are cereal grins, protein supplements (both vegetable and animal protein supplements), mineral supplements and vitamin supplements. In the normal feed formulation concentrations of energy (metabolizable energy) and protein are the primary consideration. For compounding composite poultry feeds from the standard feed ingredients like yellow maize, soyabean meal and fish meal do not need much nutritional knowledge but replacement of these feed ingredients needs thorough knowledge of quality and quantitative composition of other feed ingredients considered suitable for feeding and normally not used by the feed manufacturing companies.

Informations required for balanced feed formulation

Normally complete formula feeds are prepared for the feeding of poultry. The feeds are prepared from the mixing of low fibre feed ingredients of high digestibility and nutritive value. Following factors are considered:

1. **Class of poultry:** The composition of complete poultry feed is affected by the species, breed, strain, age, sex, level of egg production and management practices.
2. **Nutritional requirements:** Poultry required high energy-high protein compounded feeds properly balanced for the optimum supply of not only metabolizable energy and proteins but also for essential amino acids, essential fatty acids, vitamins and minerals. The levels of amino acids and minerals should not be antagonistic.
3. **Feed intake capacity of the birds:** Due to growing stage there is little scope of feeding low energy low protein diets to chicks and broilers but there is some room for the lowering of protein and metabolizable energy with in the limits of optimum nutrients supply for egg production.
4. **Selection of feed ingredients:** Following points should be kept in mind during the selection of feed ingredients for compounding of complete balanced feeds for different types of poultry.

(a) Acceptability of feeds: Spoiled feeds or compounded feeds containing damaged feeds of bad odour and taste are rejected by the birds or intake is highly decreased.

(b) Chemical composition particularly the fibre content and acid insoluble ash content in the ingredients.

(c) Presence of anti-nutritional factors like phytates, oxalates and protease inhibitors.

(d) Status of fungal damage and mycotoxins in the feed ingredients.

(e) Easy and plenty availability.

(f) Competitive price of feed ingredients for reducing cost of feeding without any adverse effect on health, growth and egg production.

(g) Feed ingredient should not impart bad odour in eggs and meat.

(h) Least requirement of special treatments for the improvement of feeding value.

(i) Special attention is required for examing the essential amino acids and essential fatty acids profiles of the feeds.

Preparation of check list of nutrients in the calculated feed formulae

Commercial poultry strains are nutritionally sensitive birds and imbalance of nutrients has significant effect on production and reproduction. Feed formulation for different classes of poultry needs special attention for the balanced supply of metabolizable energy, essential amino acids, essential fatty acids, minerals and vitamins. It is known that most of the vegetable feeds are deficient in some of the essential amino acids like lysine and methionine and several minerals. It is also known that phosphorus and several other essential minerals are only partially available from the feeds due to chelation with phytic acid because phytase enzyme necessary for the release of minerals is not synthesized in the alimentary tract of birds. This aspect is more important for the caged birds.

Availability of essential supplements

Some of the amino acids, minerals and vitamins need supplementation for balancing the compounded feed formulae. Small quantity of good quality fish meal is mixed for the balanced supply of essential amino acids. Similarly minerals and vitamins need supplementation. Among these, minerals mixture can be stored for longer period at a dry place to prevent lump formation because some of the salts used for the manufacturing of mineral mixture are hygroscopic. Fish meal and vitamin supplements require frequent fresh supply because fish meal attracts insects and microorganisms that damage the quality and vitamins loose potency after some time due to which date of manufacture, expiry date and storage conditions are prominently depicted on the vitamins package.

Steps taken before the determination of the ratio of cereal grains and protein supplements

It is now known from the detailed chemical analyses of common feeds used for the preparation of compounded feeds that certain essential amino acids are deficient. These are commonly referred as limiting amino acids. In order to minimize the problems of balancing essential amino acids a mixture of two or more proteinous feeds is prepared. Care is taken in the identification of proteinous feeds wherever it is possible.

Limiting essential amino acids in common feed ingredients for poultry

Some of the common feed ingredients used for the preparation of poultry feeds are presented in Table 11.1 alongwith the limiting amino acids in order of priority.

Table 11.1: Limiting amino acids in different feed ingredients

S.No.	Name of feed ingredients	Limiting amino acids (s)
A.	**Cereal grains or Energy feeds**	
1.	Maize or corn	Lysine, tryptophane
2.	White sorghum (Jowar)	Lysine
3.	Barley	Methionine
4.	Rice Kani/Broken rice	Methionine, cystine
5.	Small millets	Lysine, methionine, cystine
6.	Wheat	Lysine, Methionine, Tryptophane
B.	**Plant protein supplements**	
1.	Soyabean meal	Methionine
2.	Groundnut cake	Methionine, Lysine, Tryptophane
3.	Rape/ Mustard seed meal	Lysine
4.	Cotton seed meal	Lysine, Methionine
5.	Sesame (Til) meal	Lysine, Threonine

Methods of feed formulation

The common methods of hit and trial calculation, Pearson's square and algebric equations are still used. However, use of computer is increasing in feed formulation by hit and trial process and it is more useful for using several ingredients for balancing the nutrients in compounded feeds. Least cost linear programming is a complicated method of feed formulation and still not perfect.

The feed formulation methods requiring minimum mathematical involvement are more acceptable than the tedious methods.

1. **Use of Pearson's square:** In this method only one major nutrient either the level of protein or the level of energy can be worked out at a time. This method is more used for herbivorous animals in which quantity consideration is relatively more important than the quality.

   Example: For preparation of a feed mixture of 20% protein content for poultry, one ingredient should be a cereal grain and another should be a mixture of vegetable protein and fish meal.

   Procedure: The proportion of starchy cereal grains in poultry feeds is approximately 60-65% and protein supplement is 30-35%. About 7-10% fish meal is mixed in poultry feeds for balancing essential amino acids requirement.

   Suppose yellow maize with 10% protein content is used with groundnut meal of 40% protein and fish meal of 60% protein content. With such feed ingredients about 14-15% protein should come from the proteinous feeds mixture and the mixture will constitute about 40% of the compounded feed.

   Therefore, protein mixture CP = 15 x 5/2 = 75/2 % = 37.5%

   Thus, the ratio of groundnut cake and fish meal will be determined as follows:

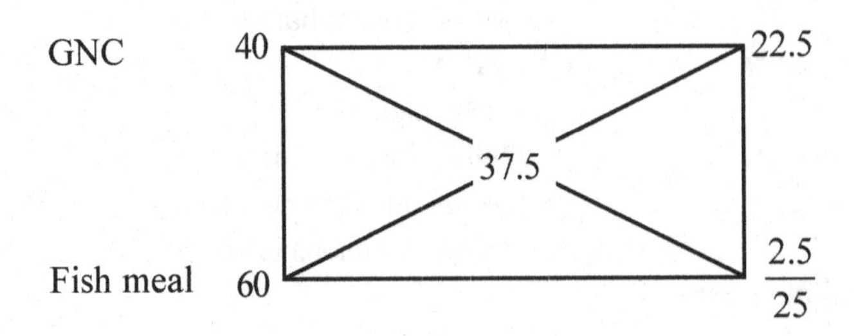

   Since 25 parts protein mixture contains 22.5 parts GNC

   100 parts will contain = 22.5 x 100/25 = 90% groundnut cake

   And 100-90 = 10% fish meal

In second step ratio of maize and protein mixture is calculated as follows:

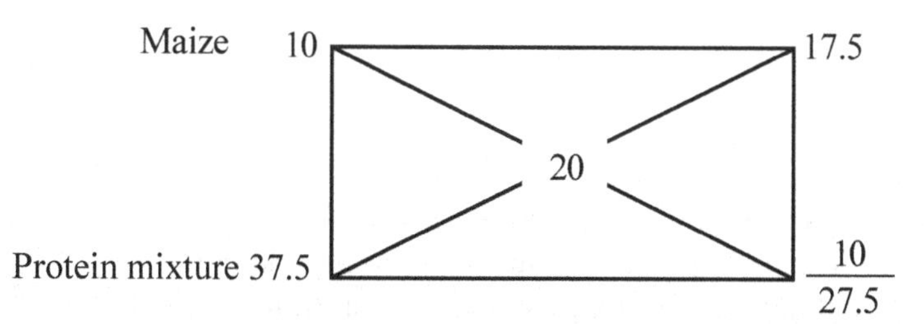

The amount of maize = 17.5 x 100/27.5 % = Maize = 63.63 % or 63.6% and protein mixture =100-63.6 = 36.4%

Therefore, final diet will contain-

Maiz 63.6%

And protein mixture = 100-63.6 = 36.4%

Ground nut cake 22.5 × 36.4/25 = 32.75%

Fish meal 36.40 – 32.75 = 3.65%

2. **Use of algebric equations:** This method is also used for the determination of ratio of only two feeds, i.e. an energy feed and a protein supplement as follows:

$A + B = 100$

$xA + yB = C$

where, A is grain or grain mixture or energy feeds and B is protein supplement. C is the % protein content in final mixture.

x is protein % in A and y is protein % in B.

Example: Compute a poultry feed of 20% Protein content from maize containing 10% protein and protein supplement of 50% protein content.

Now, $A + B = 100$ equation (i)

$.1A + .5B = 20$ equation (ii)

Multiply equation (i) with 0.1 and equation (ii) also with 0.1 to get the following:

$.1A + .1B = 10$ —(iii)

$.1A + .5B = 20$ —(iv)

Subtraction of (iii) from (iv) yields

.4 B = 10

Or B = 10/0.4 = 25%

And A =100-25 =75%

3. **Least cost feed formulation:** It is a complicated mathematical calculation for a poultry nutritionist and require computers. Linear programming, however can take care of balancing several nutrients at a time. This requires development of programmes through the use of nutritional composition of different feed ingredients. For the calculation of a balanced compounded feed mixture the informations required by the computer are the metabolizable energy and other nutrients of the bird and also the ME value, protein and other nutrients in the feed ingredients.

4. **Use of computer in hit and trial method, and development of softwares for feed formulation:** The development of feed formulation software requires informations on the nutritional requirements of poultry and the nutritional composition of feed ingredients to be used for feed formulation. A software for poultry feed formulation has been developed by a group of Animal Nutritionists of the Central Avian Research Institute, Izatnagar. It is available for the use of concerned persons in poultry production against payment.

Limits of use of feed ingredients in the compounded feeds of poultry

Standard balanced feeds of poultry are the mixture of different feed ingredients, supplements and additives. Since all kinds of feeds are not equal in chemical composition and feeding values and many contain one or more antinutritional and harmful substances. Therefore, on the basis of palatability, nutritive value, antinutritional factors and their tolerance limits maximum levels have been worked out on the basis of feeding experiments and experience.

1. **Inclusion levels of starchy (energy) feeds:** Starchy feeds specially the cereal grains constitute the major part of poultry feeds. If available at a competitive price yellow maize is the most preferred feed for poultry as stated by Late professor Subodh Kumar Talapatra over half a century back that " Poultry industry develops around yellow maize". In 1970 we were able to supply the requirement of vitamin A activity from the carotenoids of only 35% yellow maize in the diet of white leghorn layers. Over 40% yellow maize was replaced by barley (N.N. Pathak and S.K. Talapatra, 1970). The effective level of various energy feeds in the diets of poultry are presented in table 11.2.

Table 11.2: Inclusion level of starchy (energy) feeds in the compounded diets of poultry (% w/w).

Sl. No.	Feedstuff	Content in mixed feed %	Factors to be considered before use
1.	Yellow maize	70	Mycotoxins
2.	White maize	70	Mycotoxins, pigments
3.	Maize grit	10	
4.	White sorghum	50	0.25-0.35 % tannins
5.	Yellow sorghum	50	Low tannins, low carotenoids
6.	Brown sorghum	30	0.5-0.6 % tannins
7.	Red sorghum	15	1.8-2.2% tannins
8.	Peerl millet	30	Ergot alkaloids
9.	Wheat	50	Arabinoxylans, avoid new grains
10.	Discoloured wheat	20-25	Mycotoxins
11.	Damaged wheat	Upto 25	Mycotoxins level depends on damage
12.	Broken rice	Upto 50	Brown rice may be rancid
13.	Rice kani	Upto 20	Thaimine deficient
14.	Rice bran	20	Fibre, rancidity, mycotoxins
15.	Deoiled rice bran	20	Fibre, dustiness
16.	Rice polish	25	Rancidity, mycotoxins
17.	Deoiled rice polish	20	Dustiness, mycotoxins
18.	Finger millet (ragi/madua)	15	In layers, mostly escape digestion
19.	Foxtail millet (tagun)	20	High fibre, contains carotenoids
20.	Paspalum (Kodon)	15	Ergot like alkaloids
21.	Cheena	20	High fibre, contains carotenoids
22.	Barley	20	High fibre, beta-glucans
23.	Pearled (dehusked) barley	50	Beta-glucans
24.	Oats	15	High fibre, beta-glucans
25.	Tapioca chips	10	Cyanogenic glycosides
26.	Molasses	10	High potassium, hygroscopy
27.	Bakery residue	10	Mould growth, mycotoxins
28.	Vegetable oil (edible)	Upto 8	Including fat in feeds
29.	Animal fat	5	Lard and tallow are prohibited by law in Bharat

2. **Inclusion levels of proteinous feeds (% w/w):** Protenous feeds are the second major component of poultry feeds and may constitute upto 40% of mixed diets of young chicks and broilers (Table 11.3).

Table 11.3: Inclusion levels of proteinous feeds.

Sl. No.	Feedstuff	Inclusion level (% w/w)	Factors to be considered before use
1	Soyabeans	10	Protease inhibitor, mycotosins
2	Soyabean meal (Solvent extracted)	40	Mycotoxins, methionine, main aflatoxins
3.	Groundnut cake (expeller)	30	Mycotoxins, main aflatoxins
4.	Groundnut meal (solvent extracted)	25	Mycotoxins, main aflatoxins
5.	Cotton seed meal	5	Gossypol
6.	Mustard-rape seed meal (expeller)	10	Tannins, goitrogens, erusic acid
7.	Do (Solvent extracted)	10	-do-
8.	Sesame meal	15	Cyanogenic glycosides
9.	Sunflower cake	10	Chlorogenic acid, fibre
10.	Safflower cake	10	Fibbre, saponinlike bitter in oil
11.	Maize gluten	5	High methionine, deficient lysine
12.	Maize germ meal	10	
13.	Peas	10	Protease inhibitors
14.	Toasted guar meal	10	
Non conventional protein feeds			
15.	Karanj cake	10	Karanjins
16.	Kusum cake	10	
17.	Mahua seed cake	3	Mowrin
18.	Neem seed cake	3	Nimbin, Nimbidin
19.	Niger cake	5-10	
Animal Protein supplements			
20.	Fish meal	7-10	Salt, sand, urea, scales
21.	Dried whole fish	7-10	Salt, sand
22.	Dried fish solubles	4-5	Low protein
23.	Fish trash, dried	4-5	Sand, salt, scale, gut content
24.	Meat meal (sterilized)	10	Mould growth
25.	Poultry trash meal	4-5	Sterilization
26.	Hydrolyzed feather meal	2	
27.	Hatchery waste meal	3	Sterilized
28.	Shrimp trash meal	5	Sterilized

[Table Contd.

Contd. Table]

Sl. No.	Feedstuff	Inclusion level (% w/w)	Factors to be considered before use
29.	Skim milk powder	3	Sterilized
30.	Silk worm pupae meal	3	
Miscellaneous protein feeds			
31	Torula yeast	5	
32	Bakers yeast	4	
33	Dried distillers grain	10	Fibre, mould growth
34.	Pencillin mycelium waste	10	
35.	Molasses sludge (dried)	3-5	
36.	Berseem leaf meal	5-10	Carotenoids, calcium
37.	Lucerne leaf meal	5-10	Caroteniods, calcium

Note: Non conventional and miscellaneous feeds are generally used in the diets of layers and foraging/scavenging birds:

Information about minerals and vitamins supplements are given in chapter on nutrients.

12

THE RATITES (FLIGHTLESS BIRDS)

The ratites are probably the advanced stage of avian evolution exhibited by consipicuous modifications in the morphology. Some important features of ratites are:

1. These are flightless birds.
2. Keel bone is absent on the breast bone or sternum.
3. Long bones contain bone marrow and red blood cells are produced.
4. Long bones of legs and wings do not contain air cells that help in flight by reducing weight.

Some Common ratites of economic importance

The ratites are mostly heavy birds capable of producing highly nutritious meat and eggs. Feathers are the valuable by-products extensively used for the manufacture of various kinds of commercial products of artistic values. The other products are oil of medicinal and cosmetic values. Some of the species used for commercial farming are:

1. Emu farming is done for meat, eggs, oil and feathers.
2. Ostriches are the heavier ratite reared for feathers, leather, meat and eggs.
3. Rheas are the source of meat, eggs, feathers and skin etc.
4. Kiwis are still wild and now protected in New Zealand.
5. Cassovaries (Asuwarius spp.) are also ratites of Australia and islands of the continent.

The EMU

It is a useful ratite of Australia and has been introduced in many countries including India for commercial farming. At present there are more that 700 emu farms of different size in the states of Kernataka, Tamil Nadu, Kerala, Maharashtra, Gujarat, Zharkhand, Andhra Pradesh, Uttar Pradesh and Haryana. The bird has been widely accepted and will further spread in other states. The medicinal value of emu oil is becoming the main commercial product.

Zoological classification

Kingdom: Animalia

Phylum: Chordata

Class: Aves

Order: Struthioformes (or Casuriiformes)

Family: Casuaridae

Genus: Dromaius

Species: Dromaius novacholandiae

Sub-species

1. D.n. novachallanede
2. D.n. woodward
3. D.n. rathschildi
4. D.n. diemenensis

However, there are still dispute in the number of sub-species and it may take some more time to resolve the issue.

The emu is a heavy and tall non-flight bird and second in size among the living ratites of the world. The tallest ratite is its cousin, ostrich. Emus prefer to live on grasslands away from heavy human habitation, dense forests and dry deserts. The adult birds may be upto 200 cm (80 inches) tall and weigh upto 45 kg (100 lb). The long neck is pliable and sharp eyes are helpful in finding food grains, insects and larvae on the ground. The legs are long and thin. Toes are only three in each foot. The birds move at a moderate speed but may run at a speed of upto 50km per hour, if chased or otherwise. The long legs are helpful to take strides of 2.5-3.0 metres (about 8-10feet) during running. Emus travel long distance in search of foods during scarcity and may become nomadic.

The nails on toes are sharp and strong which are used during fighting and protection from the predators. The legs are very strong and can break wire fence. Sharp eye sight and auditory power help in detecting enemies from a long distance.

The feathers provide protection from ambient heat due to which emus remain active even during the mid day. The heat tolerance ability is high and thermoregulation mechanism is efficient. There is no sexual dimorphism in shape, size and plumage colour.

Breeding season is generally May and June. The emus are polygamous. Fighting among females for a male is frequently observed. The females try to attract the male by characteristic vocalization emitted by manipulating an inflatable sac in the neck. Female in oestrum mates many times with the same or different mates. She lays eggs in several batches in a season which are incubated by the father during the incubation period of 50-56 days. He looses significant weight during the incubation period due to starvation. However, both sexes put enough weight before the breeding season. The chicks are nurtured by the father for approximately six months but continue to live upto next breeding season.

Common predators: Besides humans the common predators of wild emus are the wild dogs (Dingo), large eagles and hawks. Young and active emus protect themselves by kicking the dingos but only run to protect from the bill wars of eagles and hawks. The life of emus on such ranges is 10 to 20 years.

Feeds: Although the bird is omnivorous but bulk of the diet is made of different herbages. The crop is not developed and main food holding part for digestion is the small intestine. Average length of small intestine, caeca and large intestine (colon to cloaca) is 315, 12 and 29 cm constituting about 88.5, 3.3 and 8.2 percent respectively. Microbial digestion occurs in the intestine and most of the nutrients are absorbed through the walls of the pre-cloacal compartments (small and large intestines).

Products

1. **Eggs:** Large greenish tinged eggs are nutritious like any other edible eggs of birds.
2. **Gourmet or emu meat:** Emu meat contains very low percentage of fat. It is red meat also known as" New heart healthy meat". Protein content is high and richer in many essential minerals and vitamins. Gourmet is also free from contaminants as the emus are raised on pastures away from the roads and factories.

3. **Emu oil:** It is a highly valuable product of medicinal value and also used in cosmetics.
4. **Skin and leather:** Used for preparation of items of fashion and decoration.
5. **Feathers:** These are inferior and less valuable than the ostrich's feathers. Used for the manufacture of various kinds of art and craft items.
6. **Bones:** Preparation of craft goods.
7. **Toe nails:** These are processed for the manufacture of jwelleries and decoration items.
8. **Droppings:** The excreta pellets are dried and used as fuel and also as organic manure in the crops.

Properties of emu oil

1. Emu oil has deep penetration properties and reaches the deeper tissue through skin, specially the joints.
2. It is anti-inflammatory.
3. It possesses bacteriostatic properties and prevents many bacterial diseases.
4. It is a good emulsifier.
5. It stimulates wound healing.
6. Emu oil is considered to have anti-aging property.
7. Mild irritation reaction on human skin.
8. It is non-carcinogenic.
9. Very good moisturizer.

These inherent natural properties have made emu oil very useful for humans and used for the manufacture of medicines and cosmetics.

Clinical and cosmetic uses of Emu oil

Pure emu oil is used as a constituent of several formulations used for the following purpose.

1. Application of pure emu oil provides relief from the arthritis and rheumatism.
2. Mixed with vaseline or liquid paraffin its application on the skin delays the aging changes. Its regular application increases skin thickness.

3. Scars of pimples and wound are considerably reduced by the regular application of emu oil and its clinico-cosmatic formulations.
4. Used as a base for increasing the effect of several drugs by synergistic action.
5. Pure emu oil is used topical for the cure of pulled muscle, sprain and joint pain. Therefore, very useful for sports person, persons doing heavy physical works and aged persons.

Commercial uses of feathers

The feathers of emu are inferior in quality in comparison to the feathers of ostrich. Feathers are used for the manufacture of fashion items and other useful items like duster, fan, masks, toys etc. Rough feathers are used for finishing metal articles before painting.

Characteristics of emu leather

The leather made of emu skin has high commercial value due to the following features.

1. Emu leather is thin, pliable, durable and supple.
2. Unique full quelled pattern makes attractive.
3. Different colours may be applied.
4. Working is easy with emu leather.
5. The raised imprints of feather follicles on the skin provides a simmering surface.
6. Moderate thickness and softness make emu leather suitable for clothings and accessories.
7. Leg skin is in demand for making belts and watch straps.

Uses of Emu leather

1. Manufacture of light weight coats and jackets.
2. Used for making purses and hand bags.
3. Leg skin is used for making belts and straps.
4. Fancy and decoration items.

Uses of egg shell

Large size empty egg shells are used for the manufacture of decoration articles, lampshades and jwelleries. Broken egg shells are offered to emus in their fenced paddocks.

The OSTRICH

The ostrich is a largest living bird on the earth. It belongs to ratite group. Different species are found in African and Arabian countries Farming of ostriches for meat, feathers and skin production is increasing and the bird has been taken to many countries. The running speed is highest among the long ratites like emu and rhea.

Zoological classification

Kingdom: Animalia

Phylum: Chordata

Class: Aves

Order: Struthioniformes

Family: Struthionidae

Genus: Struthio

Species; Struthio camelus

Sub species and native habitat

1. S.c. australus	South African ostrich
2. S.c. camelus	North African ostrich
3. S.c. massaicus	Massai ostrich
4. S.c. syriacus	Arbian ostrich
5. S.c. molybdophanes	Somali ostrich

Some zoologist consider Somali ostrich a separate species. However, for commercial farming these minor discripencies are non-significant.

The bird is largest among the living large ratites emus and rheas. The pliable neck and strong legs are long. Each foot contains three toes with strong nails. It is fastest running living bird and may run upto a speed of 95-100 km per hour.

The ostrich is technically omnivorous but the greater proportion of diet is made of variety of herbages comprising leaves, shoots, fruits, berries, bulbs and tubers etc. Like emu the proventriculus (true stomach) and gizzard are large for holding and grinding large quantity of foods. The crop is absent. Contrary to emu the small intestine is short and large intestine is long. The average length of small intestine, caeca and large intestine of adult ostrich are about 512, 94 and 800 cm comprising about 36, 7 and 57 percent of the total length of intestines respectively. Some insects, larvae, pupae, snails and other smaller creatures may be eaten alongwith the forages. Role of enteric microorganisms in the digestion of fibrous foods of plant origin needs detail exploration.

Reproduction is seasonal which differs according to agro-climatic conditions. Each eligible male forms a harem of two to seven breedable females in the flock. Fight for luring females is quite common. Harem size of robust and dominant males is greater than the other males. Some times fight may be fatal. Legs are frequently used for kicking. After mating the females lay eggs in the nest. Size of eggs ranges 700 to 900 g. Incubation period is about 8 weeks and chicks require parentral help in early life. However, they are left free to form their juvenile group of 10 to 50 before the on set of the next breeding season. Fidelity is not assured and females may elope from one harem to another.

Valuable products of ostrich

1. Beautiful feathers fetch high price and used for the manufacture of different kinds of variety items and decoration of garments.
2. Low fat meat has high demand.
3. Skin is processed for leather making and used for the preparation of garments, boots, sandles, purses, bags etc.
4. Infertile eggs are used for food.
5. Egg shells are used for making decoration pieces.

THE RHEA

It is a heavy ratite of South America famous for its green shaded large eggs and low fat red meat. The wild species are distributed widely on the grasslands and mountaineous regions. The main species is greater or American rhea (Rhea americana); and another Lesser or Darwins' rhea (Rhea pennata) was recognized and included recently (2008) in rhea group. Rhea has been tamed and domesticated for farming on range land in Americas and Europe.

Zoological Classification

Kingdom: Animalia

Phylum: Chordata

Class: Aves

Order: Struthioniformes (Rheiformes)

Family: Rheidae

Genus: Rhea

Species: Rhea americanea (Greater or American Rhea)

Rhea pennata (Lesser or Darwin's Rhea)

Each species has been further differentiated into several sub species on the basis of morphological differences among the rheas of different regions and flocks. Greater or American rhea possesses five and lesser or Darwin's rhea three subspecies.

Sub species of greater rhea (Rhea americana)

1. Rhea americana americana (Central and east Brazil)
2. Rhea a. albescens (Argentina)
3. Rhea a. araneipes (Brazil, Bolivia, Paraguay)
4. Rhea a. intermedia (Brazil and Urugnay)
5. Rhea a nobilis (Paraguay)

Sub species of lesser or Darwin's rhea (Rhea pennata)

1. Rhea pennata garleppi (Argentina, Bolivia, Peru)
2. Rhea p. pennata (Argentina, Chile)
3. Rhea p. tarapacensis (Chile)

More informations are available on the greater Rhea which is extensively reared on farms for the production of meat, egg and some other commercial products. Therefore, more emphasis has been given on the American rhea which is also known as common rhea or grey rhea. Some of the local names of greater rhea in different parts of the south America are ema(Portuguese), nondu (Spanish),

Choique (Mapudungum) and suri (Quenchua). The bird is similar to emu and ostrich in appearance. Normally it is not easy to identify the subspecies of rheas because they have been differentiated on small differences in morphology like the degree of black colour of the throat and place of abundance etc.

The American rhea is a tall and heavy non flying bird of 20-27kg (44-60lb) body weight, 150 cm (60iches) height and 129 cm (51 inches)long from the beak to end of the tail. Sexual dimorphism is found and males are taller and heavier than the females. Some adult males of greater rhea weigh upto 40 kg (88lb) and stand upto 183 cm or 6 feet. The neck is long and pliable which is normally carried vertical but used for picking foods from ground to shrubs and small trees.

The beak is long, eyes are sharp and head is relatively small. Long legs are strong and possess marrow but no air space. Toes are only three in each leg. The long wings are used to balance the body at sharp turns during running. The plumage are fluffy and tattered. The colour of feathers is various shades of grey and brown, but white feathered and albino specimen are also found. Plumage colour of males is normally darker than the females of same family. Plumage of newly hatched chicks is grey with dark stripes on the body along the dosrsal line.

Foods of Rheas

The rheas are omnivorous but greater proportion of diet contains variety of leaves, fruits, flowers, grasses and seeds. Foods also include insects, amphibians, reptiles, rodents small birds etc. Rheas often eat substantial quantity of coarse forage. Grits and stone pieces are engulfed to help grinding of ingested feeds in the gizzard.

Unlike other birds urine is collected in a separate sac of cloaca.

Vocalization

Only young chicks make noise in early life probably for foods from the parents. Males vocalize only during the hot months of breeding season to attract eligible females. They remain almost silent during the non-breeding season and form groups of 10 to 100 birds in a flock. Large size flocks are common in greater or American rheas.

Reproduction

1. Hot months are the season of mating that extend from October to March in different regions.

2. The males are polygamous and the females are polyandrous.
3. Males make nest on grassland during the mating season.
4. During breeding season males give call to the females for mating which takes place near the nest to facilitate egg laying.
5. Courtship of males take place with 2 to 12 females at a time and number is normally higher with the dominant males.
6. After mating, females lay eggs in the nest. The number for incubation ranges from 10 to 60 eggs of about 600g (21oz) weight and 13 cm (5.2 inches) length.
7. After laying eggs, the female couples with another male and previous male takes care and incubate the eggs.
8. Incubation period is 29 to 43 days.
9. Some dominant and intelligent males employ subordinates for the incubation and care of eggs for hatching and pair with other females for mating.
10. Hatching is completed with in 36 hours for all incubated eggs in one nest.
11. Clutch size is 5-10 eggs at a time.

Uses

1. Rearing for edible eggs and meat production.
2. Oil is used for cosmetics and soaps.
3. Plumage, feathers and leather are used commercially.

Predators

Puma, a large wild cat and humans are the known predators of wild and feral rheas.

Orphan eggs

Some times without completing laying the female runs away after mating and lay scattered eggs on the grassland. Some stray females also lay scattered eggs without mating. Such eggs are called orphan eggs. The orphan eggs may be fertile also probably due to parting with previous mate without laying all the eggs in a clutch.

Comparison of egg characteristics of domestic ratites

The ostrich, emu and rhea are the domesticated ratites and used for commercial farming in many countries. Emu farming is becoming popular in many states of India. The comparative value of some egg characteristics may be useful for developing egg carriers and also trays for incubation. Some common traits of ostrich, emu and rhea are presented in Table 12.1.

Table 12.1: Some characteristics of ratites egg.

Traits	Ostrich	Emu	Rhea
Gross weight (g)	1520±93	594±49	640±54
Egg shape index	1.23±0.03	1.45±0.08	1.55±0.10
Yolk weight (g)	331±35	226±21	180±36
Yolk as % of egg	21.8	38.0	28.1
Albumen weight (g)	893±56	282±37	381±40
Albumen % of egg weight	58.7	47.5	59.5
Albumen index	0.07±0.02	0.11±0.03	0.05±0.02
Egg shell weight with membrane (g)	296±25	86±13	79±7.5
Do as % of egg weight	19.5	14.5	12.4
Egg shell thickness (mm)	–	–	–
with membrane	2.13±0.1	1.22±0.1	1.00±0.1
without membrane	2.03±0.1	1.16±0.1	0.96±0.1

THE KIWI (Apteryx spp.)

The Kiwi is a flightless wild avian species of ratite group. It is a native of New Zealand and has been recognized as a National symbol of the country. The population of kiwi has drastically decreased due to extensive hunting and predators. Now hunting is banned and programmes have been initiated for conservation.

Zoological classification

The Kiwi is smallest surviving ratite on the earth and concentrated in the native land, the New Zealand. Its placement in animal kingdom may be depicted as follows:

Kingdom: Animalia

Phylum: Chordata

Class: Aves

Order: Struthioniformes

Family: Apterygidae

Genus: Apteryx

Species: 1. Apteryx australis (Brown kiwi)

2. A. haastii (Great spotted kiwi)

3. A. owenii (Little spotted kiwi)

4. A. montelli (North Island brown kiwi)

5. A. rowi (Okarita brown kiwi)

Apteryx = "A" means no and "teryx" means wing in greek language.

Among the five species, the greater spotted kiwi is largest and little spotted kiwi is smallest. The other three species are closer to greater spotted kiwi in size:

1. **Greater spotted kiwi (A. haastii):** Sexual diamorphism is apparent with taller (45 cm. or 18 inch) and heavier (3.3 kg or 7.3 lb) female than the male of 2.4 kg (5.3 lb) body weight. Grey-brow plumage is decorated with lighter bands. One egg is laid in a cycle and incubation is done by both parents.
2. **Little spotted kiwi (A. owenii):** is a small bird unable to protect from the predators. Average height is 25 cm (10 inches) and body weight is 1.3 kg (2.9 lb). Female lays one egg at a time that is incubated by the male parent.
3. **Brown kiwi (A. australis):** It is also known as Tokoeka and found in many areas of the south island of New Zealand. The bird is comparable with great spotted kiwi in shape and size. However, plumage colour is light brown. It has many recognized sub species.
4. **North island brown kiwi (A. mantelli):** It is most abundant among the all species. Adult female at 40 cm (16 inches) and 2.8 kg (6.2 lb) body weight is heavier than the male of about 2.2 kg (4.9 lb) body weight. Streaked red-brown plumage is more attractive. Female lays 2 eggs from a clutch which are incubated by the male parent.
5. **Okaritio brown kiwi (A. rowi):** It is also known as Okarito or Rowi. This species is slightly smaller than the great spotted. Kiwi. Plumage colour is grayish and in some birds white facial feathers may be seen. This is most prolific among the five species and female lays upto 3 eggs in different nest in a season. Incubation is done by both parents.

Some important informations about Kiwi

1. Egg size of kiwi as proportion to body size is largest among the birds. Egg size may be upto one-fourth of body weight (300g to 500g in different species).
2. These are generally monogamous couple.
3. Breeding season extends from June to March and pair calls each other in the night. Mating occurs in the nesting area at interval of three days.
4. One to three eggs are laid in different species in a season.
5. Couple kiwi has been recorded upto 20 years.
6. Both ovaries are functional but egg production is highly controlled.
7. Shy and nocternal.
8. Sense of smell is highly developed. The nostrils are situated at the end of long beak. The sense of smell is used for collecting food from deep water, mud and debris without seeing.
9. Common foods are seeds, fruits, worms, small cray fish, eels, small amphibians and their larvae etc.
10. Eggs are smooth and ivory white or greenish in colour.
11. Incubation period is 63 to 92 days in different species.
12. In the formation of huge egg demand of nutrients supply increases for which female is required to eat 3-4 times more than the normal diet of non-breeding period.
13. The egg occupies most of the abdominal space pushing aside the alimentary canal and female is forced to fast upto laying.

Special morphological features

1. Keel bone is absent on the sternum which is required to anchor the wing muscles.
2. The wings have been modified to minute vestiges.
3. The modified feathers are bristled like coarse hairs and forked.
4. The bones are not hollow and pneumatic but contains marrow as in mammals.
5. Preen gland present in most of the other ratites is absent.
6. Bill is long, pliable and sensitive to touch, which makes it suitable for food exploration.
7. Eyes pectein is reduced.
8. Kiwi possess 13 modified flight feathers, no tail but a small pygostyle. The feathers do not contain barbules and aftershafts. Large vibrissae are found around the gape.
9. The gizzard is weak but capacious, caecum is long and narrow.

13

EGGS AND MEAT FOR FOOD

Egg production is a reproductive function of female and starts with the onset of puberty. Fertilization of egg is essential for the formation of zygote and development of embryo to full form representing the species. The importance of poultry egg as a highly nutritious food for human has been well recognized and now only infertile eggs are produced for table purpose. The chicken and other poultry birds have been developed in many high egg producing strains through the use of modern breeding (cross breeding) and selection technologies. The largest quantity of table eggs of poultry are produced by the chicken followed by duck and other birds in India. A small quantity of table eggs of indigenous local fowl is still produced and preferred by many Indians but largest number is counting from the commercial layer farms of different high laying strains of leghorn (mostly white leghorn) birds. Now chicken strains producing more than 300 eggs annually are quite common and researches are on for the development of strains of still higher production. Taking about 26 horns for the complete formation of eggs from ovulation (rupture of mature follicle) to oviposition (expulsion) of complete egg there is scope of obtaining 337 eggs in a year (365 days).

Now chicken production has been differentiated in to two major groups of laying strains and fast growing broiler strains.

Laying strains

These are slow growing chicken attaining puberty to start continuous cyclic laying at about 22-24 weeks of age. These birds are further divided into two categories for the specific production purpose.

1. Production of fertile eggs to be hatched for the productions of laying hens (replacement flock).
2. Production of infertile eggs exclusively for table purpose (food).

Broiler strains

These are synthetic strains of heavy and medium size chicken breeds evolved for very high growth rate, meat quality and feed conversion efficiency. Broiler strains are fed at two different rates for differential growth.

1. Feeding of low energy diet for slow growth to start laying at 22-24 weeks of age.
2. Feeding of high energy and high protein balanced diets for finishing at 5 to 7 weeks of age and 1 to 2 kg or even higher body weight.

Egg production in chicken

Egg production is a normal reproductive function of female poultry bird and chicken has been evolved as a super layer. The reproductive tract or egg formation apparatus of a hen is formed of a pair of ovaries and female reproductive tract comprising of infundibulum or funnel of oviduct, magnum, isthmus and opening of vagina into cloaca (Fig. 13.1).

Function of ovaries

1. Development of oocytes to form follicles.
2. Maturation of ovarian follicles.
3. Rupture of mature follicle and release of ovum.
4. Synthesis and release of steroids is influenced by the growth (size) of follicle. Largest follicle releases progesterone and also some amount of testosterone and oestradial. The second largest follicle in growth sequence produces largest amount of testosterone while third largest follicle produces oestradiol.

   Steroids in ovarian follicles are secreted by two different types of cells. Progestorone is secreated by granulosa cells and, testrotrone and oestrsdiol are secreted by the theca interna cells.

Structure of Ovary

The ovary is made of outer cortex layer formed of oogonia and the inner medulla consisting of vascular, neural and connective tissue. All nutrients for the development and maturation of follicles are supplied by the vascular net work of medulla. A functional ovary appears like an irregular mass containing 5-6 or more developing follicles of different size.

Only left ovary and oviduct are functional in chicken and other poultry birds. Occurrence of functional right ovary and right oviduct is very rare, and occurrence of both ovaries and oviduct in functional form is a rarest of rare incidence. The functional left ovary is attached closely on the surface of dorsal abdominal wall just in front of the left kidney and posterior to the left lung along the posterior vena cava.

A functional ovary of adult chicken is quite large in relation to body weight. The ovary of a chicken hen of about 2kg body weight may weigh 45g (40-50g) including 5-6 g weight of maturing 5-6 follicles in progressive sequence. Maturing of the follicles is a continuous and sequential process which is controlled by photoperiodism in almost all avian species. A functional ovary of chicken contains about 12000 oogonia arranged in the medulla. However, like any other species (mammals and reptiles) only small fraction gets opportunity for development and oviposition in the reproductive life of birds.

An immature ovarian follicle is oocyst enclosed by follicular cells of medulla. The constituents of an immature follicle are the nucleus and watery or serous follicular fluid containing other organelles enclosed in a strong cell membrane. The progressive growth of many follicles occurs in sequence to the stage of maturity. Ones the larger follicles bulg out prominently on the surface of ovary, these are called Graafian follicles. A fully mature follicle differs conspicuously from the developing microscopic follicles. The size of mature follicle is very large due to formation of yolk. The yolk along with ovum is covered with a proteinous membrane known as the perivitelline membrane which is considered to be formed by the follicular cells encircling the yolk and ovum. Nucleus is placed on the top of yolk. The perivitelline membrane is garlanded by a layer of granulosa cells with several projections penetrating into yolk. The stratum granulosa is covered by the rich vascularized and innervated thecal layers. Steroids synthesis occurs in granulosa cells from factors in theca interna.

Venous venules rich complex vascular system is also formed with the fast developing follicles and prominant larger veins are easily seen on the surface of larger follicles. Stigma is a band on the surface of follicles which is almost avascular and fragile. On complete follicular development the stigma ruptures from the pressure of follicular fluid and oocyte comes out of the follicle.

The translucent grape skin like debris left on the surface of ovary after the escape of oocyte and associated contents still functions for guiding transport of ovum towards oviduct through infundibulum. After the rupture of follicles the spot becomes physiologically non functional as contrary to mammals there is no formation of corpus luteum. Probably this is the reason for almost continuous oogenesis in selectively bred domesticated poultry birds.

Oogenesis

At the time of hatching formation of germinal cells on the ovaries is not yet complete and takes further 3-4 days for the development of germinal cells to complete oogenesis. With the progressive growth one of the overy (mostly left ovary) grows and with the maturity of ovary some time before the onset of puberty formation of white round disk takes place beneath the vitelline membrane. This disk is known as blastodisk and it does not develop further in the birds kept unbred for table egg production. The process of egg formation continues without fertilization and normal oviposition occurs on complete formation of egg in about 25 hours passage through the genital tract.

The colour, size, shape and appearance of visible follicles change with the progressive development and deposition of nutrients in different forms. The smallest ovarian follicle is watery white. In the intermediary follicles of different size colour of content gradually changes to yellow and intensity of yellow colour increases with the increase in the size of follicles. Appearance of yellow colour (or dull colour in absence of dietary yellow pigments as on feeding of white sorghum, wheat, barley or rice) appears in the follicular contents. In poultry the process of yolk deposition in early phase of development is slow and takes about 2 months. At this stage size of visible yellow follicles is 5mm to 8mm. After this, growth of few follicles is suddenly stimulated for high growth and follicles attain 30-40mm size in 8-11 days. The phenomenon of identification of growing follicles for very fast growth is probably not yet known. Likewise, rejection, atresia and regression of some other follicles is also not known. These changes in follicular development starts in growing phase and completed just before the onset of puberty exhibited by sexual behaviour and first oviposition at about 22 weeks of age in domestic chicken (Gallus domesticus).

Structure of oviduct

The oviduct of a poultry including chicken is a tubular structure starting as infundibulum near the ovary and ending in the cloaca. The entire structure is made of internal layer of tunica mucosa layered by tunica muscularis comprising of inner circular and outer longitudinal muscle fibres and covered by a layer of tunica serosa. The oviduct is a highly tortuous tubular organ measuring 60 to 80 cm long on extending after removal from an adult hen.

The oviduct is generally differentiated into five regions on the basis of functions from ovarian end to opening in the cloaca. These are infundibulum or funnel, magnum, isthmus, uterus and vagina. The vagina opens in to the cloaca which is common passage for the expulsion of digestive wastes, urinary wastes and the reproductive products. Each of the five regions are distinct from each other and perform specific function in egg formation (Fig. 13.1).

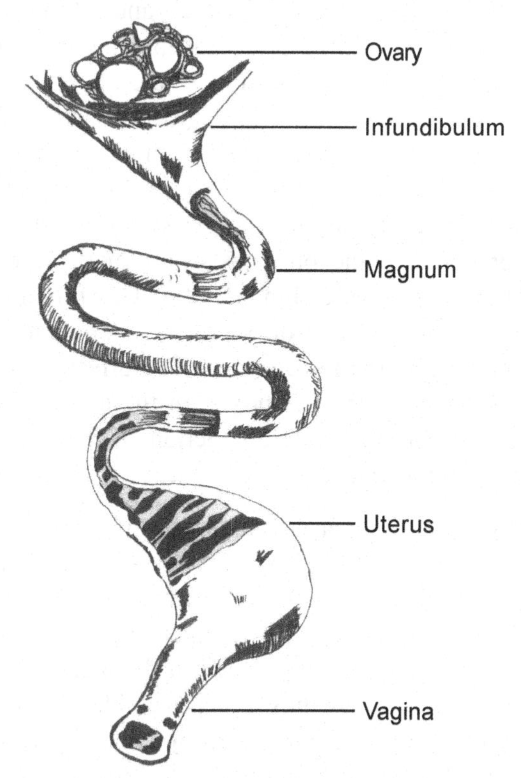

Fig. 13.1: Genital tract of hen.

The infundibulum is a funnel shaped structure of 7-8 cm length made for receiving and transporting ovum through the oviduct. The anterior part of infundibulum is membranous with few muscle fibers which gradually become heavier posteriorly due to increase of muscles fibers and glands. The change from posterior infundibulum to anterior magnum is also gradual. Only small portion of posterior infundibulum is glandular. The glands are similar to the glands of magnum. The infundibulum is highly contractile at the time of approaching ovulation and engulfs the follicle before ovulation to swallow the ovum with release. There is no scope of ovum dropping in the abdominal cavity. Unlike mammals the infundibulum in avian species does not possess fimbriae for the passage of ovum into the oviduct. In the inseminated birds fertilization occurs just at the time of entrance into the infundibulum before the start of deposition of albumen around the yolk.

Although secretion of albumen or egg white begins in the distal glandular portion of the posterior regions of the infundibulum but profuse secretion occurs in the magnum. Magnum is largest part and forms almost half of the length of oviduct (30-35cm). The wall of magnum is thick and whitish. The mucosa of magnum contains large number of tubular glands opening on the luminal surface. The mucosa is made of ciliated cells and non ciliated goblet cells. Albumen is composed of several proteins of different biological properties. Moreover, main function of albumen is the protein reserve for development of embryo (in fertile egg). Albumen secretion in magnum is controlled by the coordinated action of oestrogen, testosterone and progesteron. The growth of oviduct is regulated by oestregen but function requires the other two hormones. Goblet cells secrete special protein known as avidin which binds biotin. This function of goblet cells is regulated by progesterone. The process of development in magnum has synergistic effect of oestrogen and testosterone. The effect of progesterone may be either synergistic or antagonistic depending on the dose. However, these are experimental findings. In normal conditions action of these three hormones is complex and not yet fully illustrated.

The ovum covered with albumen layer moves to isthmus which is 8-10cm long and separated from the magnum by a short non-glandular visible band. Two keratinous, fibrous shell membranes are secreted by the isthmus to envelope the egg contents entered from the magnum. The outér membrane provides base for shell formation by the secretion of shell glands in the uterus. The 9-12 cm long uterus is posterior continuation of isthmus. It is pouch like, thick walled muscular sac possessing longitudinal mucosal folds lined by ciliated and non-ciliated cells and tubular glands beneath the epithelium. For the formation of egg shell the calcium and bicarbonates are primarily secreted by the ciliated cells and the tubular glands respectively.

A constricting muscular band known as utero-vaginal sphincter controls the passage connecting uterus with vagina. Some tubular glands in the vagina (9-12 cm long) provide medium for long duration storage of live spermatozoa in the bred hens. The vagina is last segment of tract that joins cloaca. The egg is fully formed before entering into the vagina where it is not retained and expelled out within a minute. In breeding hens vagina is a place of storage of spermatozoa that are capable to migrate to upper part of infundibulum to fertilize ova for several days. The functions of different segments of oviduct and approximate time required in each segment are summarized in Table 13.1.

Table13.1: Functions of different segments of oviduct in egg formation in hen (Gallus domesticus).

Segment of oviduct	Main Functions	Duration
1. Infundibulum	To engulf ovum and fertilization in mated birds.	15 minutes
2. Magnum	Secretion of albumen	180 minutes
3. Isthmus	Secretion of shell membranes	110 minutes
4. Uterus (shell gland segment)	Addition of fluid to egg content, stratification of albumen, formation of chalazae (twisted albumen), shell production and secretion of shell pigment in some breeds.	1200 minutes (20 hours)
5. Vagina	Storage of sperm (in mated hens); expulsion of egg.	1 minute
	Total time (approximate)	1506 minute (25 hours)

Function of oviduct and utreus: The functions of a functional oviduct after puberty may be counted as follows:

1. To catch ovum from the ovary at the time of ovulation.
2. Fertilization of ovum in the upper part of the infundibulum in mated hens.
3. Step wise downward progressive transportation of ovum during formation through different successive segments.
4. Deposition of albumen layers around yolk and ovum in magnum.
5. Secretion of sell membranes around egg contents in isthmus.
6. Addition of fluid in the egg contents specially the albumen. The process is called plumping.
7. Stratification of albumen into four distinct layers.

(i) Twisted albumen band from either side of yolk to polls. These strands are called chalazae. Inner albumen layer surrounds the yolk and chalazae are extension of inner layer.

Chalazae is considered to hold yolk and also embryo in fertile eggs in position. In fertile eggs hatchability of eggs with higher ratio of thick albuman is higher.

(ii) Next is a thin layer of watery albumen.

(iii) Next is a layer of thick albumen and

(iv) Last is again a layer of thin albumen.

8. Secretion of shell occurs in uterus by the shell glands. Largest part of shell is formed of calcium carbonate (about 98%) bound by a glycoprotein matrix (about 2%). The crystalline part of egg shell formed of columns of material embedded in the outer shell membrane. These columns are separated by fine pores extending from outside of egg to shell membrane for facilitating gas exchange by embryo in fertile eggs.
9. An Outer thin proteinous layer on the egg shell is cuticle which probably prevents bacterial invasion and helps in increasing self life.
10. Pigments are impregnated in egg shells of birds producing pigments in uterus. This is a breed and species trait. The colour of egg shell is characteristic not only for species but also for the breeds and strains. The shell colour can be also changed by selective breeding as it has been done at the Central Avian Research Institute, Izatnagar in Japanese quail (Cuturnix cuturnix). The original colour of egg shell of imported Japanese quail is patched brown painted. Now white egg shell strains has been evolved from the same stock through selective breeding.

Oviposition

The expulsion of eggs out of the body through vent is known as oviposition. Although formation of the eggs does not change the behaviour of layers but the birds sit silently in the laying nest or its cage for few minutes before the expulsion of eggs. The sitting duration before oviposition ranges from 2-3 minutes to 10-15 minutes in different birds. Non commercial local native poultry birds including fowl examine the egg immediately after oviposition and in such birds oviposition occurs on the bed of dry grass or other material and kept protected. This shows the high mother instinct of native fowl in which laying is an exclusive reproductive function for the multiplication of species. This hereditary characteristic of poultry has disappeared in the commercial high egg producing strains. This change is probably due to significant change in laying pattern from limited egg laying only in breeding season synchronized by natural clock to almost daily laying. The function has rather become a routine like that of excreta droppings and rarely any commercial bird bather for the safety and protection of egg after laying. Such birds are not disturbed by the collection of eggs from the place of laying. It is rather refractory.

Egg

The egg of avian and reptiles is a special nutritional arrangement provided by the nature in the process of evolution for the nourishment of developing embryo outside the body of mother. Protective arrangements are perfect both for physical protection by a hard calcareous egg shell and nourishment is provided by a mixture of essential nutrients contained in yolk and albumen in sufficient quantity to meet the requirements during development in incubation period.

Poultry egg

Great variation occurs in the size and shape of eggs of different avian species. For the description chicken egg has been used as an example or representative because consumption of chicken eggs is much higher than the eggs of any other avian species. In fertile eggs all the nutritional requirements of developing embryo is supplied upto 48 hours post hatching. During this period chicks learn to identify and eat feeds. In many avian species newly hatched chicks are unable to eat themselves and fed by mother for several days or weeks as in the case of pigeon and sparrow etc.

Physical composition of egg

An egg is formed of egg shell, a pair of shell membranes, albumen, yolk and embryo (only in fertile eggs).

1. **Egg shell:** It is hard outer cover of the egg and constitute about 10 (8-11) percent of the total egg mass. The shell is a strong structure consisting of about 94percent calcium carbonate, 1 percent each of magnesium carbonate,

calcium phosphate and 4 percent organic matter (mostly protein forming the matrix of egg shell). The shell is a porous structure for the exchange of air in a fertile egg with developing embryo.

The surface of egg shell is covered by a thin mucous membrane known as cuticle which protects shell pores but allow air exchange.

2. **Shell membranes:** The outer shell membrane lies on the inner surface of shell and the inner one encloses the egg contents.
3. **Albumen or egg white:** It is a watery substance formed of five different viscosity, viz. outer thin albumen, middle thin albumen, inner thin albumen, inner thick albumen and a chalazae the twisted thick albumen. The chalazae hold the yolk in position.
4. **Vitelline membrane:** It is thin, transparent membrane which encloses yolk and does not permit mixing with albumen.
5. **Egg yolk:** It is more or less spherical and formed of several concentric layers deposited during the development of follicles on the ovary. Yolk constitutes about 30(27-32) percent of the total egg. Generally proportion of yolk content is higher in smaller size normal eggs.
6. **Germ disc or Blastoderm:** It is clearly formed in the fertile egg. A passage connecting blastoderm with centre of the yolk is called latebra.
7. **Air cell:** An air cell is present on the larger end of egg. It is formed due to cooling of egg causing contraction of inner fluid contents after oviposition. The size of air cell is variable and increases on aging. An air cell in fresh egg is smallest and increases with the increase in storage days. Grading of table egg quality is done by candling of eggs by an experienced technical hand. The grades are AA for fresh eggs containing smallest air cell, A for moderate old but good for table purpose. B grade is more older and should be subjected to floating test in water. A floating egg should not be eaten and it should also not be used for the feeding of pets and other animals.

Shapes of chicken egg

Shape of egg is important for arrangement in trays for incubation of fertile eggs as well as transportation of fertile and table eggs. Shape of an egg is presented as shape index which is calculated as follows.

$$\text{Shape index} = \frac{\text{Maximum breadth of egg}}{\text{Length of egg}} \times 100$$

Shape index of normal eggs of chicken ranges from 60 to 80.

Size of chicken eggs

Size or weight of eggs is an important trait that determines the suitability for the incubation of fertile eggs and food value of table eggs. The eggs of uniform size normal for a breed are most suitable for setting in the trays of incubator. The table eggs are shorted in different grades on the basis of weight for marketing. Grading system of table eggs varies among the countries. In the Bhartiya markets grading of table eggs is not yet popular.

Grading of eggs

The objectives of egg grading are different for the fertile eggs for incubation and table eggs for eating.

Fertile eggs are separated on the basis of weight into four groups of extra large, large, medium and small. Only eggs of standard shape are selected for incubation. Small eggs and that with abnormal shape, soiled, thin shell and large air cell are not used for incubation. The other factors considered for the selection of fertile eggs are the internal contents like firmness of albumen, position of yolk, size of air cell, blood spot and meat spot in internal contents are detected by candling.

The table eggs are first separated in two groups, A and B on the basis of candling. All eggs with blood spot, meat spot and large air cell (about 50% or more larger than the air cell of fresh egg) are segregated in B category and spot less clean eggs are kept. in A category (Table 13.2). The eggs of abnormal and deformed shapes and, thin shell are also excluded from A class.

Table 13.2: Grading of eggs as per Agmark standards in Bharat.

Grade	Weight (g)	Shell	Air cell	White	Yolk
I A grade					
A Extra large	60 g and above	clean, unbroken sound normal Shape	up to 4mm in depth, practically regular or better	clear, reasonably firm	Fairly well centered practically free from defects, outline indistinct.
A large	53-59				
A Medium	45-52				
A Small	38-44				
II B Grade					
B Extra Large	60g and above	clean to moderately stained, sound and slightly Abnormal	8mm in depth, may be free and slightly bubbling	clear, may be slightly weak	May be slightly off centered, Outline slightly visible
B Large	53-59				
B Medium	45-52				
B Small	38-44				

Cleaning of eggs

After collection dirty (stained and soiled) eggs are sorted. Only sound eggs are cleaned with soap and water or any other detergent recommended for public use. Temperature of water used for cleaning is kept higher than the temperature of the egg. Egg washing machines are used for cleaning large number.

Packing of eggs

Packing of the eggs depends on the distance and means of transportation. Following factors are considered for the selection of egg packing containers:

1. It should provide maximum cushioning effect for minimum breakage during handling and transportation.
2. Each tray should contain equal number of eggs to facilitate counting.
3. Egg containers should be low cost, durable and easy to clean.

Egg containers are made by handmade rough papers, pulp of recycled papers, card board and plastic. Size of tray ranges to carry 6 to 24 eggs for domestic uses and 36 or more eggs for transportation. The trays are arranged in rows and columns in the transportation containers. Total weight of an egg container is normally kept 10 or 20kg for easy handling during loading and unloading of the consignment.

Eggs are normally transported on head, cycle, rickshaw and auto-rickshaw in local market; auto-rickshaw and mini-truck for short distance; and truck and rail for long distance. Other facilities like boat and steamer are also used on availability.

Modified method of egg grading for commercial marketing of table eggs

In addition to freshness and cleanliness the consumers of egg are also concerned with the size that determines the quantity of edible contents and availability of nutrients.

The additional measures of egg grading earlier recommended by Agmark include the gross weight of per unit of 10 eggs or a dozen of eggs in nearest gramme (g) unit as shown in Table 13.3.

Although, grades of table eggs have been decided long back but hardly grading is used in practice. Probably grades of table eggs are taken into consideration by the large confectionary units utilizing egg contents on quantitative basis. General consumers are satisfied with selected clean and larger eggs from the available lot of the retail shops.

Table 13.3: Parameters of table egg grading in Bharat.

Grade	Weight of egg(g)			Egg	Air	Egg	Yellow
	Single	unit of 10	Per dozen	Shell	Cell	White	or Yolk
A quality or fresh eggs							
A Extra Large	60 & above	596 & above	715 & above	clear, un broken and sound, shape normal	up to 4mm in depth, practi-cally regular or better	Clear, reaso-nably firm	fairly well centered Practically free from defects, outline indistinct
A Large	53-59	526-595	631-714	do	do	do	do
A Medium	45-52	446-525	535-630	do	do	do	do
A Small	38-44	380-445	456-530	do	do	do	do
B Quality or preserved eggs							
B Extra Large	60 & above	596 & above	715 & above	Clean to modera-tely stained and sound, shape, slightly abnormal	8 mm in depth, may be free and slightly bubbly	Clean, may be slightly week	May be slightly off centered. Outline slightly visible
B Large	53-59	526-695	631-714	do	do	do	do
B Medium	45-52	446-525	535-630	do	do	do	do
B Small	38-44	380-445	456-530	do	do	do	do

Edible Contents in standard eggs of different avian species

The average size of standard eggs and percentage of different contents are presented in Table 13.4.

Table 13.4: Size and egg contents (% on fresh basis)

Species	Egg weight (g)	Albumen (%)	Yolk (%)	Shell (%)
Chicken	58	56	32	12
Duck	80	57	33	10
Goose	200	53	35	12

[Table Contd.

Contd. Table]

Species	Egg weight (g)	Albumen (%)	Yolk (%)	Shell (%)
Guinea fowl	40	52	35	13
Turkey	85	56	32	12
Quail	10	47	32	21
Pigeon	17	71	20	9
Emu	595	47	38	15
Ostrich	1520	58	22	20
Rhea	640	60	28	12

Gross Nutritional composition of different eggs

Avian and some of reptile eggs (turtle and tortoise) are very rich sources of highly digestible nutrients. Mankind is using eggs of many avian species and few reptile species for eating since time immemorial. Nutritional composition of eggs of same common avian species is presented in Table 13.5.

Table 13.5: Nutritional composition of egg contents of different avian species (%) and energy (Keal) Value.

Species	Moisture	Protein	Lipids	CHO	Ash	Energy
Chicken	74	13	12	1	1	178
Guinea fowl	73	13	12	1	1	178
Turkey	73	13	12	–	2	178
Quail	74	13	11	1	1	170
Pigeon	73	14	11	1	1	175
Duck	71	13	14	1	1	195
Goose	70	14	13	1.6	1.4	190
Emu	67	12	16	–	5	120

Amino acids composition of table eggs of chicken and other avian species

Amino acids concentration specially that of essential amino acids determine the protein quality of avian eggs. The eggs are considered as one of the richer sources of essential amino acids comparable with milk, meat, and fish etc. A comparative chart of amino acids composition of two avian species is presented in table 13.6.

Table 13.6: Amino acids composition of table eggs of chicken and other avian species.

Amino acids	Chicken	Emu
A. Essential amino acids for human		
1. Isulucine	5.43	5.6
2. Leucine	8.53	8.3
3. Lysine	7.19	9.7
4. Methionine	3.09	3.8
5. Phenylalanine	5.35	5.5
6. Threonine	4.76	5.1
7. Tryptophane	1.25	—
8. Histidine (for children)	2.34	3.0
9. Valine	6.1	6.5
B. Semi essential amino acids		
10. Arginine	6.02	6.5
11. Cystine	2.3	—
12. Tyrosine	4.10	6.1
C. Other amino acids		
13. Aspartic acid	10.03	8.4
14. Alanine	5.60	5.7
15. Glutamic acid	13.04	12.3
16. Glycine	3.34	4.1
17. Proline	4.01	—
18. Serine	7.44	9.7

Fatty acid composition and Cholesterol Content of egg

Almost all lipids in an egg is present in the yolk. Egg yolk lipids can be separated into the following three groups of constituents of nutritional importance for humans.

1. Unsaturated fatty acids 70% approx in chiken egg
2. Saturated fatty acids 28% approx.
3. Fat soluble constituents 2% approx

   (Lecithin, carotenoids, sterols)

Fatty acids composition of chicken egg has been compared with the egg fats of Emu bird in table 13.7.

Table 13.7: Fatty acids composition of egg liquids (%)

Fatty acids	Chicken	Emu
A. **Saturated fatty acids**		
C 14:0 Myristic acid	0.47	0.2
C 16:0 Palmitic acid	26.98	25.5
C 18:0 Stearic acid	9.59	5.4
B. **Mono unsaturated fatty acids (MUFA)**		
C 16:1 Palmitoleic acid	3.07	4.7
C 18:1 Oleic acid	42.40	53.2
C 20:1 Gadoleic acid	0.25	0.1
C. **Poly unsaturated fatty acids (PUFA)**		
C 18:2 Linoleic acid (EFA)	13.00	8.8
C 18:3 Alfa-linolenic acid (EFA)	0.35	0.2
C 18:4 Moroctic acid	0.12	—
C 20:4/6 Arachidonic acid	1.65	1.0
C 20:5 Tinnodonic acid	1.06	—
C 22:6 Docosahexaenoic acid	0.59	—

EFA = Essential fatty acid

Cholesterol content in egg liquids

Dietary cholesterol is invariably blamed for causing cardiovascular disorders in humans. Cholesterol is present in all foods of animal origin and egg is one of the richer sources of cholesterol.

Preservation of table eggs

Preservation of eggs is done for prolonging the self life, keeping quality and food value of the table eggs. In areas of large size layers farm daily disposal of table eggs may not be possible. Similarly medium and small poultry farmers do not produce enough eggs to justify the rent of transportation. Under such situations preservation of eggs is required. It is also required in remote areas lacking facilities of regular supply of small quantity. The following factors should be taken into consideration for preservation of table eggs.

1. It should be natural clean because cleaning removes the protective cuticle from the surface of egg.
2. Eggs should be intact without any crack in shell.

3. Egg shell should be strong and without any deformity .
4. If preservation of washed eggs is necessary they should be dried well before preservation.
5. Wet eggs should never be preserved because scope of microbial invasion causing spoilage of contents increases due to percolation of infected water. Two types of spoilage are found in the preserved wet eggs. These are green rot spoilage caused by Pseudomonas group and black rot spoilage by Proteus group of bacteria.

Principles of preservation of whole eggs

1. Prevention of microbial invasion by sealing the pores of egg shell.
2. Retardation of microbial proliferation, if entered by any means.
3. Reduction of evaporation loss and escape of gases.
4. Delaying unwanted physico-chemical changes in egg contents.
5. Defertilization of fertile eggs.

Method of preservation of shell eggs

Various methods of shell egg preservation are either used alone or two or more methods are used simultaneously.

1. **Thermal preservation:** Three processes used for thermal preservation of eggs are (a) flash thermal treatment, (b) thermo stabilization or regular thermal treatment and (c) thermo stabilization and simultaneous cooling.
   a. **Flash thermal treatment:** The eggs are immersed only for 2-3 seconds in hot water (71-100°C) and taken out and dried at room temperature. In this process harmful microorganisms on the shell surface are destroyed and coagulation of a thin layer of outer albumen forms another barrior. In this process minimum changes occur in the physical form of the egg contents.
   b. **Thermo stabilization or regular thermal treatment:** In this process eggs are immersed in hot water for longer period depending on temperature of water used. Duration of immersion at 49°C (120°F), 54°C (129°F), 56°C (133°F) and 60°C (140°F) is 35, 15, 10 and 5 minutes respectively.
   c. **Thermo stabilization with simultaneous oil coating:** The two process complement each other in prolonging maintenance of quality for longer period. Thermo stabilization process 1.b and 1.c coagulates the thick albumen and fresh appearance of eggs is significantly prolonged.

2. **Low temperature preservation:** The eggs are kept at a cool place for about 3, 9, 25, 65 or 100 days at 37°C (99°F), 24°C (75°F), 16°C (61°F), 7°C (45°F) and 1°C (33.8°F) respectively.

   Immersion of eggs in liquids: This statement needs minor modification for depicting the liquid used. Actually different aquous solutions contain lime, sodium silicate, borax, chlorinated water, caustic potash solution, caustic soda solution, magnesium oxide solution and magnesium sulphate solution. Therefore, it will be more appropriate to say immersion of eggs in aquous mixtures or water mixtures. The common immersion methods used in aquous mixtures are (a) water glass method and (b) lime water or lime sealing method.

   a. **Water glass method of egg immersion:** The water glass mixture is prepared by mixing equal amount of sodium silicate and water. This is stock solution and 5percent of this stock solution in water is used for preservation of eggs by liquid immersion method. The eggs are immersed in this diluted solution and left over night. During immersion period a thin layer of silicate is deposited on the surface of eggs. The silicate is antiseptic and does not alter odour, flavors or taste of the egg contents.

   b. **Lime water immersion or lime sealing:** Basal solution is made by mixing 1kg of milk lime with one kg of hot water (may be boiling) in a strong vessel. The mixture is cooled to room temperature and the liquid strained from the mixture is diluted with 4-5 liters of cold water and 225g finely powered common salt is added, mixed and allowed settling. After settling of un-dissolved constituents the supernatant is decanted in a suitable vessel and used for the immersion of eggs for preservation . The eggs are immersed in lime water for 16-18 hours, taken out and air dried at room temperature and then arranged in egg trays (filter flats) and stored at room temperature(20°C-25°C). These eggs can be preserved for 3-4 weeks at room temperature. The alkalinity of lime imparts preservation effect to greater extant by forming a thin layer on the egg shell. The thin film of calcium carbonate has considerable sealing effect of pores.

3. **Oil treatment:** Two methods of oil treatment of eggs are used for short and long duration preservation. The thin layer of oil formed around the egg shell seals the pores stopping the exchange of air and entrance of damaging microorganisms.

   a. **Oil spray treatment:** In this process oil is sprayed with the help of a sprayer on the eggs while turning mildly. In this process oil film formation may not be complete. Such eggs can be preserved for few days.

b. **Oil immersion process:** In this process egg shell is properly coated and pores are sealed. The clean laid eggs are dipped in oil with in few hours for preserving the internal contents intact. Washing of eggs with hot water for 3-4 seconds and oil coating produce better preservation effect.

4. **Cold storage:** This method is generally used in cooperative areas where eggs are first collected from the small holding producers and then transported to dealers for marketing. The eggs are stored at 12.5°C-15.5°C (55°-60°F) and 70-80% relative humidity (RH). For longer storage temperature is brought down to 10°C (50°F) and 80-90% RH is maintained for minimizing evaporative loss of moisture of eggs.

5. **Combination of oil coating and cold storage:** Oil coated fresh eggs can be preserved for longer duration as shown in Table 13.8

Table 13.8: Preservation duration of oil coated and uncoated eggs at different temperature and RH in cold storage.

Conditions	Temperature (°C and ° F)	RH(%)	Preservation duration
Oil coated eggs			
1	14°C (57.2°F)	90	240 days (8 months)
2	13.3°C (56°F)	85	60 days (2 months)
3	20°C (68°F)	72	28 days (about 1 month)
Unoil coated eggs			
1	14°C (57.2°F)	90	180 days (6 months)
2	13.3°C (56°F)	85	30 days (1 month)
3	20°C (68°F)	72	7 days (1 weak)

6. **Use of gases for egg preservations:** This is possible only when air tight recepticles are available for egg storage in the environment of carbon dioxide. It provides greater protection to consistency of albumen. Maintenance of 1.55 ppm ozone in storage containers helps in controlling fungal growth and development of off odour.

7. **Use of earthen pots or pitcherss:** This is also a low temperature preservation method. Ordinary unglazed earthen pitchers are placed in wet sand exposed to natural air blowing. This has been found to keep internal temperature of pot below 30°C even during the summer. Internal temperature will fall further in cold season. Small flock holder egg producers use this method for few days for sale in the weekly market of the area or saling to dealers.

8. **Salting of eggs for preservation:** The principle of egg preservation is the ex-osmosis. Salt is hygroscopic and common salt is a cheap and easily available edible product. For salting of eggs saturated salt solution (brine), pulverized salt or salt-clay mixture paste are used. The high concentration of salt draws water from bacteria and moulds due to which they are dehydrated, become inactive and loose the ability of multiplication. Thus, microbial spoilage of eggs is controlled.

   a. **Salting of eggs in brine:** Generally earthen pitcher is used for egg salting. Drinking water is taken in an earthen pitcher and common salt is dissolved to make saturated. It is determined by some residue of non-dissolved salt in the brine. The fresh laid eggs are cleaned and placed in the brine.

   b. **Salt-mud plaster:** In the process a thick paste of equal amount of pulverized salt and clay are taken and made into a thick paste by adding water. Now fresh laid clean (or cleaned) eggs are rolled in the salt mud for deposition of a reasonably thick plaster.

   c. **Storage in salt powder:** The fresh laid eggs are placed covered with salt powder and it was used by people working in salt making during the olden days. Now it is not used.

Preservation method of Century egg

The life of table egg is prolonged for longer period ranging from several weeks to many months. The duration of preservation depends on the quality of eggs used for preservation and the efficiency of use of the processing technology.

Materials used for the coating /preservation of century egg

Sieved ash of plants (dry grasses, fuel wood or charcoat, clay (Yellowish or black or any other as per availability), powdered common salt, powdered lime and fine chaffed as well as long paddy straw.

Method of coating

An easily smearable paste is prepared by mixing fine ground clay, fine ground lime and fine ground common salt. After this 10-15 percent fine chaffed paddy straw is mixed with the salt mixture. Now a reasonably thick paste of mixture is prepared by mixing adequate water. The salt-clay paste should be thick enough to stay on the surface of egg shell after plastering. The salt-clay plastered eggs are dried and then stored at a suitable place on a thick bed of long paddy straw.

Changes in egg contents during preservation

Significant changes in the colour, consistency, odour and flavor of egg contents occur during the processing and preservation. The yolk turns dark green and cream like. Strong odour is produced by metabolic changes in protein and sulphur amino acids. Albumen turns in the form of a dark brown transparent jelly. The chemical changes are considered due to alkaline substances in the egg. The pH of preserved eggs may be highly alkaline (pH 9-12). The century eggs can be consumed generally by the people accustomed to eat such preserved eggs emitting strong odour of ammonia and sulpher gases.

Preservation of liquid egg

Dehydrated as well as liquid whole egg, albumen or yolk are used in different bakeries, confectionary and other products. The liquid egg contents, mixed or separate are preserved either frozen or dehydrated.

1. **Freezing of liquid egg contents:** The liquid egg content collected from fresh and clean/cleaned eggs are collected and mixed (whole egg content) or separated into albumen and yolk in different containers. The contents of different containers of whole egg content, albumen and yolk are mixed independently without contaminating each other. Shell membrane, if comes in albumen and vitalline membrane are removed. After this three types of liquid egg contents are frozen quickly by an air blast process at 1.1°C (34°F). The products are labeled and stored below-32°C. Frozen yolk is prepared as plain yolk, salted yolk and sweetened (sugar, honey mixed yolk) and yolk emulsion. Freezing of whole egg contents or yolk causes gelation which is much higher in frozen yolk.

   For reducing gelation freezing and thawing combination is used at a very fast speed. Gelation of egg yolk can be controlled by mixing either 10 percent common salt or 10 percent table sugar but the use of such products will be limited to the production of only food items. Mixing of small amount of pepsin (0.04%) in egg yolk before freezing prevents gelation without any adverse effect on the utility of frozen yolk.

2. **Dehydration or drying of egg contents:** Egg powder is prepared by processing eggs in the following sequence.

   (i) Whole eggs are cleaned in hot water (40°C) containing 2 percent hypochlorite for 3 minutes.

   (ii) Cleaned eggs are broken and contents are collected in a sterilized vessel.

(iii) Now liquid egg contents are churned and filtered.

(iv) Pasteurized at 62.5°C for 3.5 minutes. A plate heat exchanger is used for pasteurization.

(v) Spray dried at an inlet temperature 160°C and outlet temperature 60°C. Atomizer works at 20000 r.p.m.

(vi) Product is cooled to room temperature for storage and uses.

Composition of whole egg powder made of chicken eggs

Composition of whole egg contents powder made of chicken eggs is presented in Table 13.9 as per the Beureau of Indian standards (15:4723:1968).

Table 13.9: Composition of chicken egg powder

Attributes	Values
Moisture %, maximum	2.0
Egg protein (NX 5.68)%, maximum	45.0
Lecithin and fat %, Maximum	38.0
Solubility, % by weight, minimum	80.0
pH, Maximum	7.9
Oxygen content by Weight. %, maximum	2.0
Bacteria count/g, maximum	75, 000
Yeast and mould count /g , maximum	100
Coliform count/g, maximum	100

Factors affecting shape of eggs

The shape of egg is a heridatory trait and should remain stable but in actual formation, it is affected by several factors having adverse effects on the functions of liver, ovary and oviduct. These organs contribute significantly in the formation of eggs. Some of the factors affecting the shape of eggs or actually the shape of egg shells are:

1. Coccidiosis
2. Ranikhet disease
3. Irregular supply of Calcium and other minerals in the diets.
4. Inflammation of the oviduct.

5. Copper deficiency
6. Potassium deficiency
7. Magnesium deficiency
8. Manganese deficiency
9. Iodine deficiency
10. Vitamin D deficiency
11. Choline deficiency

Abnormalities of eggs

Many times laying of abnormal eggs is also recorded on well managed poultry farms. The reasons for such sporadic occurrence are generally obscure. Some of the abnormalities recorded in chicken eggs are:

1. High variation in size ranging from less than 10g to more than 90g. However, occurrence of these extremes is rare.
2. Laying of shell less eggs is due to spontaneous and temporary failure of shell formation.
3. Double yolk eggs is also rare. It occurs due to simultaneous rupture of two mature follicles.
4. Laying of peewy eggs: These are tiny and apparently well formed eggs. These are considered to be the last laid egg of a longer clutch size. This may be also a follow up rupture of follicle after the rolling of double yolked egg. However, this view needs confirmation by egg laying recording.
5. Yolkless eggs: These are also small size eggs some times recorded in intensive farming system.

Colours of egg shell

Colour of egg shell is characteristic for different avian species and breeds. The most common egg shell colour is white and plenty in chicken followed by brown egg shell of different shade. The other colours are pinkish brown and speckled. The normal colour of shell in native Japenese quail is speckled brownish–blue-green but at Central Avian Research Institute, Izatnagar a new/stable strain of Japanese quail has been evolved that lay white eggs.

The chicken breeds with white colour ear lobes lay white shell eggs and chicken with red colour ear lobes lay brown colour eggs except Darking and Sussex breeds of English class that lay tinted/speckled eggs.

The egg shell colour of Emu is dark green and shade varies among different layers across the shell.

Raw egg should not be eaten

Avian egg contains an enzyme avidin which inhibits the digestion and availability of protein. The avidin is present in albumen and it is destroyed by heat treatment like boiling, poaching and cooking, Egg yolk is almost free from the avidin and often used for the preparation of many recipe particularly deserts like fruit yolk, ice-cream and pudding etc.

MEAT OF POULTRY AND OTHER BIRDS

Meat of poultry and many other birds are preferred by considerable proportion of meat consuming population in different parts of India and other countries. Main meat producing domesticated birds in India are the chicken broilers, duck broilers, chicken, duck, geese, guinea fowl, quail, pigeon and turkey. In recent years emu has been introduced in some parts of Bhartiya territory and in near future other ratites may be imported because these are heavy and foraging birds.

Birds used for meat production

Different species of domesticated and wild birds are used for meat production. Wide variation is observed in the preference of birds and level of finishing. In many parts of the country. Considerable number of people still prefer local chicken and ducks reared on foraging. Although hunting is banned but illegal hunting of wild birds particularly partridges, pigeon, doves, jungle fowl, water fowl, water crow, bagedi, hariyal, bustard and large varieties of birds is not unknown in remote areas of many regions. The birds reared for meat production are listed as follows:

1. Chicken

(i) Broilers finished at 5-6 weeks of age.

(ii) Cockerls or males of layer strains are finished at 3-4 months of age and raised on foraging in rural areas. The male chicks of laying strains at large poultry farms are mostly destroyed or sold or given free to small holders.

(iii) Spent birds: These are laying hens of 18-24 months of age culled after uneconomical egg production. A small number of cocks are also used at the end of active breeding life.

2. **Ducks:** Mixed sex duckling are mostly reared on foraging and scavenging in water sources, damp area along water sources and paddy fields following harvesting. At some places broilers are also produced.

   (i) Broiler ducks are finished at 6-8 weeks of age.

   (ii) Drakes are disposed at 8-12 weeks.

   (iii) Spent ducks are disposed at the end of active laying period at 18-24 months.

3. **Quail:**

   (i) Broiler quail are finished at 5-6 weeks of age.

   (ii) Spent quail hens are disposed at 15-18 months.

4. **Geese:**

   (i) Ganders are finished at 8-12 weeks.

   (ii) Spent geese are disposed at 18-24 months.

5. **Turkey:**

   (i) Poults are finished at 12-24 weeks.

   (ii) Spent turkey after 24 months of age.

6. **Guinea fowl:**

   (i) 10-12 weeks

   (ii) Spent birds at 24-30 months of age.

7. **Pigeon:**

   (i) Squabs are finished at 5-6 weeks.

   (ii) Adults are disposed at any age.

8. **Emu:** 3-6 months tentative.

9. **Ostrich:** 3-6 months tentative.

10. **Rhea:** 3-5 months tentative.

11. **Kiwi:** Now it is protected ratite.

Abbatoirs

The size, equipments and appliances of abbatoirs for poultry and other birds depends on the size, and number of the birds slaughtered on a day. Elaborate packaging is not required for disposal in local markets. Additional facilities are required for packaging at big abbatoirs processing poultry carcasses for marketing at long distance.

Transportation of birds to abbatoirs

Transportation of finished birds for slaughter at abbatoirs depends on the number and size of birds to be slaughtered. Various means of transportation of birds and their carcasses may be listed as follows:

1. Head load, 20-30 kg for short distance in remote areas.
2. Baggage sling is used for carrying 30-50 kg up to a distance of 10 km in remote areas.
3. Bicycle is used for carrying birds and carcases up to 20 km.
4. Auto or oil driven vehicles of different carrying capacity.
5. Animal power driven carts.
6. Railways.
7. Boats and steamers.
8. Aeroplanes.

Lairage

The birds brought to abbatoirs are provided comfortable housing with running drinking water facilities for rest. Feed is not offered to birds for 12 Hrs. before the sacrifice.

Ante mortem inspection

The inspection of birds for soundness is known as ante-mortem inspection. The inspection is carried by a qualified veterinarian specially trained for the job. Ante-mortem inspection of birds helps in discarding sick, aged and emaciated birds from the flock brought for slaughter. Ante-mortem examination includes observations of general appearance, temperature, pulse and respiration. Abnormal colour of comb and fracture of bones etc.

The birds isolated or ear marked for removal from the flock for slaughter are differentiated into the following groups:

1. The birds suffering from incurable diseases, severe traumatic injury particularly of leg bone, wing bone and keel and also heat stroke are declared unfit for human consumption and recommended for destruction.
2. Sick birds suffering from curable diseases at the time of inspection are isolated from the flock. Such birds may be accepted after proper treatment and recovery from the sickness.
3. Healthy birds with minor apparent injury are slaughtered separate. Fitness for human food is given only after finding fit on post-mortem examination.
4. Suspected for sub clinical sickness in which symptoms are not yet detectable on ante-mortem inspection. Suspected birds are also slaughtered separate and declared fit only after finding healthy on thorugh post-mortem inspection.

Commonly detectable diseases on ante-mortem inspection

Pyrexia, respiratory ailments, neural disorders, severe injuries and diarrhea are easily detected on ante-mortem inspection.

Slaughter of the birds

Slaughter of food animals is a complex subject linked with many Bhartiya socio-cultural taboos. However, this is gradually disappearing and now a days it is limited to small number. Main purpose of the selection of a suitable slaughter method is to ensure minimum pain and maximum bleeding. This improves cass quality. Some of the following methods are used for slaughter :

1. By breaking blood vessels and nerves at the level of atlas or atlas and axis with feast. It is used only for few birds (generally one and maximum 10.
2. Slitting the blood vessels below ear lobes but leaving the bones and nerves intact.
3. Stunning with 90 volt current and then immediately bleeding is done by slitting jugular vessels behind head.

   Incomplete bleeding is not desired. It reduces carcass quality and keeping quality.

Scalding

Properly bled birds are transferred to scalders after the ceizure of reflexes. Scalding is done at different temperature of water in the scalders and water temperature depends on the age and types of birds. Common terms used for scalding at different temperatures of scalding water are hard scalding (82-86°C), sub scalding (58-60°C), semi-scalding (50-55°C) and soft scalding (52-55.5°C).

Sub scalding water temperature (58-60°C) has been found most suitable for the defeathering of chicken carcasses. But, this method is not preferred because muscles become a little tough for cutting. Soft scalding is extensively used in modern abbatoirs for poultry.

Feather plucking or Defeathering of carcass

Before the development of defeathering machine feathers were removed manually by hand plucking. It is still used when few birds are processed at home, farm house, picnic or remote places for consumption. This method is also used by retail poultry shop which is common in the markets of Indian subcontinent, south-east Asian countries and many other countries.

In commercial houses feather plucking is done by electrical defeathering machine. Different types of feather pluckers are now available in the market. The principle of feather plucking machine operation is the rotation of rubber (now other flexible materials are also used) fingers fixed on drums which rotate in opposite direction. The birds are immediately transferred from sealder to plucker. The plucking is completed in about 2 minutes. Any feather left on the carcass is hand plucked during the removal from the feather plucker.

After feather plucking many filaplumes on the carcass look undesired and repulsive for the consumers. These hair like filaplumes are completely removed by singeing which is done by gas burners or other devices. This is followed by washing.

Now completely defeathered carcasses are placed on a sloppy clean plate-form and spray washed with forced water sprayer. Carcasses are frequently turned during washing either manually or mechanically.

Carcass dressing

This includes removal of head and legs below the hock joint or tarso-metatrtarsal joint. The oil glands present around the cloaca are carefully removed.

After this carcasses are hanged with hooks placed at neck. Now a transverse cut is given on the abdomen carefully avoiding the puncturing of gastro-intestinal tract. At the neck a slit is given and oesophagus is freed from the neck. At the lower end a circular incision is given around the cloaca. Now the carcasses are ready for post-mortem inspection by a meat inspector.

Removal and processing of giblets

Giblets include pericardium removed heart, gall bladder removed liver and gizzard cleaned of internal contents and the fibrous serous lining. The giblets are cleaned, washed and then wrapped in wax paper or any suitable wrapper along with the neck. This small package is placed in the abdominal cavity of eviscerated carcass during final packaging.

Chilling of dressed carcass

Dressed carcasses are put in cold chilling material like crushed ice, chilled water below 4°C, slush ice, ice flaeks or solid carbon dioxide. Duration depends on the carcass size and chilling materials. About 40-50 minutes are required to bring down the temperature between 2-4°C inside the muscles. The chilled carcasses are hanged in cold chamber below 4°C for the drainage of adhered water.

Packaging of carcass

The clean carcasses are packed individually or in cluster of 5, 10, 20 or 6 and 12 or otherwise decided by the trader for different types of retail shops. Various non-toxic and easily degradable eco friendly. Materials are used for the packing of carcasses.

Use of harmful and non-degradable materials like aluminum foil and polyethylene (plastic) bags or sheet have been discontinued and also banned in many countries.

Post-mortem examination of carcasses

Post-mortem examination is carried out after washing and opening of abdominal cavity containing intact internal organs. Each carcass is thoroughly inspected for the abnormal changes, inflammations and tumerous growth etc. Carcass inspection starts from the head and ends at the vent.

The organs requiring more attention are the naso-pharyngeal orifice, lymph nodes, liver, kidneys, other glands, heart and lungs including air sacs. During inspection whole or affected part of carcass is condemned and declared unfit for human consumption.

The condemned carcasses and parts of carcasses are further inspected for their fitness for the manufacture of meat meal for animals. However, this exercise is carried only in the big poultry processing industries handling large number of birds at a time.

Poultry carcass quality or Poultry meat quality assessment

Three methods of meat quality assessment are:

1. Physical characteristics like appearance, wholesomness, texture and meat:bone ratio.
2. Chemical characteristics, viz proximate composition amino acids composition and fatty acids composition. At some places minerals and vitamins are also considered. Now a days special consideration is given for cholesterol and unsaturated fatty, acids content.
3. Organoleptic characteristics include flavour, juiciness and tenderness.

Factors affecting poultry meat quality

1. **Tearing and disfiguration:** Damage caused to skin, muscles and bones during defective transportation, fighting in the flock or defective feather plucking in machine.
2. **Incomplete bleeding:** Appearance of incomplete bled carcasses is not attractive due to presence of scattered blood clots on the carcass. Keeping quality is significantly reduced by microbial invasion as blood is a very good medium for bacterial growth.
3. **Improper scalding:** Hard scalding temperature causes partial cooking of skin. This favours bacterial invasion and spoilage of meat.
4. **Freezer burn:** This term is used for the excessive dehydrated spots in the carcass. The discoloured circular spots appear on the skin around the feather follicles and irregular spots on the other parts of skin. This can be prevented by use of suitable packing material and fast freezing.
5. **Off flavour:** It is highly undesirable and reduces market value. The causes of off flavour may be bad storage causing rancidity of meat lipids, decomposition of carcass, contamination and absorption of bad odours from the surroundings as fat has high affinity for odourous substances.

Grading of dressed chicken

Bureau of Indian Standards (BIS) has prescribed criteria for the grading of dressed chicken (Table 13.10). Properly packed chicken meat can be stored for a week in freeze at 2°C but for longer storage deep freezing is essential.

Table 13.10: Characteristics of grade 1 and 2 chicken.

Grade 1 Chicken Conformation	Grade 2 chicken
1. Free of deformation that detract from its appearance or that affect the normal distribution of flesh. Slight deformities such as slightly curved or dented bones and slightly curved back may be present.	1 Slight abnormalities such as dented, curved or crooked breast, croocked back or misshaped legs or wings which do not materially affect the distribution of flesh or the appearance of carcass or part.
2. **Fleshing**	
The breast is moderately long and deep, and has sufficient flesh to give it a rounded appearance with the flesh carrying well up to the crest of the breast bone along its entire length.	The breast has a substantial covering of flesh with the flesh carrying upto the crest of the breast bone sufficiently to prevent a thin appearance.
3. **Fat covering**	
The fat is so distributed that there is a noticeable layer of fat in the skin in the areas between the heavy feather tracts.	The fat under the skin is sufficient to prevent a distinct appearance of the flesh through the skin, specially on the breast and legs.
4. **Defeathering**	
Free of pin feathers, diminutive feathers and hairs which are visible to the inspector or grader.	Not more than an occasional protruding pin feather or diminutive feathers shall be in evidence under a careful examination.
5. **Cuts and tears**	
Free of cuts and tears on the breast and legs.	The carcass may have very few cuts and tears.
6. **Discolouration**	
Free from discolouratuion due to bruising; free of clots, flesh bruises and discolouration of the skin such as 'blue back' is not permitted on the breast or legs of the carcass or on these individual parts and only lightly shaded discolouration are permitted elsewhere.	Discolouration due to bruising; free of clots, moderate areas of discolouration due to bruises in the skin or flesh and moderately shaded discolouration of the skin such as 'blue back' are permitted.
7. **Freeze burn**	
May have an occasional pock make due to drying of the inner layer of skin (derma) provided that non exceed the area of a circle 0.5 cm in diameter on Chickens.	May have a few pock marks due to drying of the inner layer of skin (derma), provided that no area exceeds that of a circle of 1.5 cm in diameter.

Preservation of poultry meat

The process of conservation of edible meat for use during scarcity, transportation and long duration storage is known as preservation.

Objectives of preservation

1. To stop physio-chemical changes by inactivation of enzymic and other reactions in the meat.
2. To prevent microbial invasion and decomposition caused by microorganisms.
3. To maintain palatability of the preserved meat.
4. To maintain nutritive value of the meat.

Methods of preservation

1. Refrigeration for short duration (2-4°C).
2. Deep freezing and storage at -20°C for very long period. The quality of frozen chicken has been assessed excellent, good and satisfactory on deep freezing for 12, 15 and 18 months respectively. However, storage duration is much shorter for the meat of duck, turkey and other birds.
3. Drying: some methods used are sun drying, oven drying, drum drying and freeze drying. Cooked poultry meat is better preserved than the raw meat.
4. Curing and smoking: This includes salting meat with a mixture of common salt and one or more glutamates , ascorbic acid, acetic-acid and phosphates at 2-4° C in a cold room. This is followed by pickling by stitch pumping or immersion in curing solution. Smoking is done for the improvement of flavour and colour of the meat.
5. Canning: Preservation by heat treatment of meat in a hermitically sealed metallic (tin) containers.
6. Radiation preservation is not preferred .
7. Use of antibiotics is almost prohibited.

Use of tenderizing agents

The meat of aged culled and spent birds is tough and requires very long time for cooking. Some herbal products and synthetic proteolytic enzymes have been found to improve the tenderness and reduce the cooking time of such meats. Tenderizing agents are injected in raw meat or in chicken before sacrifice or during cooking. Herbal products like pieces of raw green papaya fruit or green fruit of cucumber is a satisfactory tenderizing agent and added during cooking.

Carcass cuts

Poultry carcass is dissected into seven retail cuts that are 2 legs or drumstick, breast, wings, back and, neck and giblets. The drumstick is most preferred piece of poultry meat followed by breast and wings. The last two cuts, i.e back and, neck and giblets are the last to be picked from the serving bowel on the dining table. In standard menu only drumstick, breast and wings are served.

Gross chemical composition of meat of poultry and some other birds

Average values of gross chemical composition and calorie (energy) contents of some birds used for human consumption are presented in Table 13.11. The values reveal very high fat percentage in the edible meats of geese, duck and pigeon, followed by chicken, quail and guinea fowl. Lowest fat is found in the meat of turkey and ratites.

Table 13.11: Average chemical composition of the edible meat of poultry and other birds (%).

Species	Moisture	Dry Matter	Protein	Lipids	Ash	Calorie (Kcal/100g)
Chicken	68.0	32.4	20.0	11.0	1.0	206
Duck	53.7	46.3	19.9	25.6	0.8	350
Quail	73.5	26.5	20.5	5.0	1.0	150
Guinea fowl	70.2	29.8	21.0	7.8	1.0	170
Goose	51.0	49.0	16.4	31.5	1.1	385
Turkey	75.1	24.9	22.3	1.6	1.0	125
Pigeon	56.7	43.3	18.5	23.8	1.0	305
Dove	58.0	42.0	19.0	22.0	1.0	305
Emu	74.0	26.0	23.3	1.7	1.0	130

Nutritional composition of meat of birds

The average values of fatty acids, cholesterol, minerals and vitamins in the edible flesh of poultry and ratites are presented in Table 13.12.

Table13.12: Average values of protein, fatty acids, minerals and vitamins per 100g meat.

Nutrients Per 100g meat	Chicken Broiler	Turkey	Pigeon	Emu
Protein (g)	23.1	22.3	18.5	28.4
Lipids (g)	1.2	1.6	23.8	4.7
Saturated fat (g)	0.3	0.6	8.4	2.5
Mono unsaturated fat (g)	0.3	0.7	9.7	
Poly unsaturated fat (g)	0.3	0.4	3.1	
Cholesterol (mg)	64.0	73.0	95.0	87.0
Minerals content:				
Calcium (mg)	11.0	12.0	12.0	
Phosphorus (mg)		162.0	248.0	269.0
Magnesium (mg)		21.0	22.0	29.0
Sodium (mg)	65.0	61.0	54.0	65.0
Potassium (mg)		302.0	200.0	375.0
Iron (mg)	0.7	1.44	3.54	5.0
Copper (mg)			0.44	0.24
Mangansese (mg)			0.02	0.03
Zinc (mg)		1.33	2.20	4.60
Selemium (mg)			13.30	44.00
Vitamins				
Thiamin (mg)	–	0.13	0.21	0.32
Riboflavin (mg)	–	0.32	0.22	0.55
Niacin (mg)	–	0.11	6.05	8.90
Vitamin B6 (mg)	–	0.13	0.41	0.83
Vitamin B12 (mg)	–	–	0.40	8.50
Folic acid (mg)	–	4.00	16.00	9.00
Vitamin C (mg)	–	5.70	5.20	–

14

RED JUNGLE FOWL

Wild species of many birds were tamed by the humans in different countries. Limited informations available in the literatures are mostly scattered and lack systematic presentation to meet the academic norms. Poultry sciences in now an independent subject for almost half country. The two full fledged independent institutions of Bharat are the Central Avian Research Institute (ICAR-CARI) at Izatnagar and project Directorate on Poultry (ICAR-PDP) at Hydrabad. Inspite extensive development on avian species academic focus remained limited to chicken, Japanese quail and Duck. Sporadic attempts have been made on the study of other birds flying and purching in the human habitation. Some species of these birds were tamed and conditioned for solitary living in cages. These include parrots, myana, bulbul and partridge. Pigeons are reared in groups and provided housing facility in pigeon holds. Mostly made up of wooden planks. Now domestication of many such species has been banned in Bharat due to which researchers are now keeping away from the studies on the behabivour and habits of such species. In order to maintain link with such avian species available informations on some of such species are presented in this part of the revised edition.

The avian species included in this part are the Red Jungle Fowl (Gallus gallus), domesticated fowl (Gallus domesticus), Japanese ceruail (Cutur nix caturnix), Pigeon (Columba livia), parrots of 3 families and 5 sub families, partridge myana (bulbul and pea fowl).

Two important species of wet lands are swan (Cygnus sp.) and Saras crane (Grus antigone) with several sub species. The dem and of Turkey for table purpose during some festivals has increased. Hence, brief informations on some breeds of turkey (melliagris sp.) has been also incorporated.

RED JUNGLE FOWL (Gallus gallus)

Bharat is the home land of red jungle fowl (Gallus gallus). Even today one can get opportunity to see Red Jungle fowl while moving across the forest during dawn and dusk or even during other hours of the day light.

Some habitats in Bharat

Different varieties of the Red Jungle Fowl are found in the forests of Bharat from Himalaya to coastal forests and also the islands of the country. Some of the conspicuous habitats in Rhartiya territory are the Rajaji National Park, Corbett National Park, Dudhwa National Park, Manas National Park, Kaziranga National Park, Keoladeva Bharatpur bird century, Bandhavagash National Park, ramnagar National Park, Periyar wildlife sanctuary, Kishanpur sanctuary, Mahabir Swamy sanctuary and many others covering the entire country.

Morphology of the Red Jungle Fowl

This is a large bird of 60-70 cm length with distinct sexual dimorphism. The male birds are larger, heavier and possess long and coloured sickles. The tail of cocks measures 25-30 cm. The body is decorated with 14 feathers of red, brown, golden, orange, mehroon, green and grey feathers. Mixed colour feathers are more common probably due to uncontrolled and stray breeding among the families.

Sub species/varieties

On the basis of morphological variations the red jungle fowl of Bharat and other Asian countries are distinguished into different sub species or varieties (Table).

Table: Subspecies or varieties of Red Jungle Fowl.

Sl.	Sub-species	Native Land
1.	Gallus gallus gallus	Thailand,
2.	Gallus g. Murghi	Bharat, Mayammar, Bhutan, Bangladesh
3.	Gallus g. gaborillei	S.R. Vietnam, Combodian, Laos
4.	Gallus g. Bovkiwa	Java
5.	Gallus gallus sp.	Mayammar and adjoining Thailand
6.	Gallus g. giganticus	Brahma fowl of Assam (Bharat)
7.	Gallus g. domesticus	Domesticated fowl

Sexual maturity

The cocks of Red jungle fowl mature earlier than the females of the same flock. Average age at maturity of cocks is about 5 months and that of hens is 5-6 months. Breeding season extends in hot months of summer and spring. A hen lays one egg daily and clutch size is formed of 5-7 eggs. Hens are good mothers and take care of chicks during early life. Feathering of chicks in completed in 4-5 weeks. The light brown eggs are laid on a dry grain bed protected from the predators. The hens are excellent brooder. Like most of the birds the eligible cocks give characteristic calls to hens during the breeding season.

Diets

The red jungle fowl is an omnivorous bird. Common diet contains fallen grains, tender grasses, insects, larvae and pupae of insects, beetles, worms and small snails.

Habitat

Dense bushes and tall grasses are preferred habitats because these provide more protection from the predaters. Birds prefer high dry lands away from the dampness. Hens rearing chicks avoid adventurous movement during the rearing period of 3-4 months and trained chicks for feed identification, feeding and foraging. Solitary fowls may be seen during passing through the forest during dawn and dusk.

BRAMHA VARIETY OF RED JUNGLE FOWL

It is a very heavy large size coloured variety of Red Jungle Fowl of the forests along the Brahmaputra river of the north-east states of the Bharat. The bird appears to be extinct from the home land due to extensive and selective poaching. The main causes of its extinction were affinity for the human habitation and docile nature. It was easy to capture wild Brahma and tame as ornamental bird before 1947 when tress passing through their habitats was uncommon and least disturbing due to movements of motor vehicles and railway trains.

A smaller variety of Brahman was also in the chitgong hills extending upto Bangladesh. This variety was probably evolved due to spring of small size birds by the poachers and environmental factors. Smaller birds were probably considered to be youngs.

Morphology of the Brahma fowl

Brahma is the heaviest and largest variety of the Red Jungle Fowl of the east of the Bhartiya sub continent. Its territory extended from north-east states of Bharat to adjoining Bhutan, Bangladesh and Mayanmar. The body of bird is almost entirely covered with plumage. The two basic colours of feather are lights and darks. The plumage of light colour Brahma is white. Hackle feathers are black and white. The main feathers of tail are black but side hangers are white. The bird appears wider due to extended side hangers. Dark colour Brahmas are more attractive and feathers colour is silvery red and black as seen in partridges. Feathering on the abdominal surface, tail and legs is black. Many varieties of feather colour has been developed in the modern domesticated Brahma. The specimens of gold, black, blue, white, buff, Columbian and blue Columbian are available with the breeders in England. Small variety of Brahma in same colours is also available.

Production performance of Brahma fowl

The Brahma in a slow maturing fowl. Large variety of Brahma may continue to grow upto 2 years of age. In those days small flocks of large variety were reared in remote areas of upper Assam for meat production. The performance of Brahma under standard management has been presented in table.

Table: Body weight of large Brahma fowl in optimum management.

Attributes	Mean kg	Body weight lb
Cock	5.5	12
Hen	4.1	9
Cocurel	4.6	10
Pullet	3.6	8

Some significant events

1. Brahma was exported to united kingdom during 1840-1850.
2. It was introduced in USA in the beginning of nineteenth century.
3. The Brahma breeders of USA presented some birds to Queen Victoria.
4. Brahma was entered in the first book of poultry standards during 1865.

5. It is a relative (cousin) of Red Jungle Fowl of Bharat.
6. It is heaviest breed of Native (Wild) fowl.

   Initially it was a good layer but selective breeding for variety of coloured plumage has significantly decreased egg laying potential.
7. At present it appears to be extinct in homeland.

Future of Brahma in home land

As on today not only the people but also the poultry specialists of the home land are not aware that there is a breed of fowl named Brahma, a native of Assam and adjoining areas. In the interest of conservation of the native poultry germplasm it is duty to import, multiply the conserve the varieties of Brahma. An extensive search in the remote human colonies along the mighty Brahmaputra river and its tributaries some specimens of Brahma may be traced. In case o non availability in the home land, breedable birds should be imported from the breeders of U.K. and USA for breeding and conservation of the bird which was once upon a time native of this land.

15

DOMESTIC FOWL (*Gallus domesticus*)

Domestic chicken (*Gallus domesticus*) is one of the extensively studied avian species for the production of highly nutritious proteinous foods in the form of eggs and flesh. At present several kinds of layers producing more than 300 eggs annually and broiler strains finishing in 5-6 weeks are available in almost all countries. In the modern Bharat even strict vegetarian families have permitted egg eating to younger generation. Economic and food value of chicken is the principal part of this book. In this chapter biological aspect of the bird has been presented.

Zoological classification

The chicken belongs to order galleniferae in the animal kingdom as systematically presented as follows.

Kingdom	:	Animalia
Phylum	:	Chordata
Class	:	Aves
Order	:	Galleniferae
Family	:	Gallanae
Genus	:	Gallus
Species	:	Gallus domesticus or Gallus gallus dometicus

Other characteristics

Anatomy	Sexual dimorphism. Males are heavier and possess sicklas and spur
Rectal temperature	39.5 to 40°C or 103 -105°F

Domestic Fowl (*Gallus domesticus*)

Respiration rate per minute	12 to 30
Clutch size	1 to 10 eggs per cycle
Annual egg production	(a) Native hen 30 to 80
	(b) Developed hen 125-300

Production types		
(a) Egg Type breeds	-	White leg horn, Brown leg horn
(b) Dual purpose	-	Rhode Island Red (RIR)
(C) Meat type	-	Plymouth rock

Note:- Now Synthetic strains of layers and broilers are available in wide variety.

Adult body weight of non synthetic straws.

(a) White egg type fowl 1.5 to 3.0 kg

(b) Brown egg type fowl 2.0 to 4.0 kg

(c) Broiler strains 3.0 to 5.5 kg

(Now newer strains of smaller size and low fat are also available). Development of newer strains is a continuous process. There are claims for higher content of some essential amino acids.

Breeding age- 5-6 months

Breeding life- 18 to 24 months

Life span- Generally birds are not allowed to continue beyond 5 years. Most of them are finished at 3 years.

Photoperiod- About 14 hours light is required for optimum performance.

Daily feed requirement –

(a) Laying strains 110-130 g

(b) Broiler strains 120-150 g

Daily drinking water requirement -200 -300 ml.

16

PEA FOWL (*Pavo* spp.)

The peafowl is a large size terrestrial bird of clear sexual dimorphism. Two common varieties are decorated blue, the common bird of the Bhartiya sub continent and the green feathered variety. Pea fowls are also found in the countries of Indo-Pacific region. The bird belongs to pheasant group. Sometimes albino or albinoid freaks are also found. Indian peacock is famous for extravagant trains of feathers of the pseudotail, which they spread like broad flag during dancing. Peafowl has been introduced in many countries as ornamental bird. In many countries as ornamental bird. In the Indian sub continent peafowl is a protected bird and feathers of peacock are used for the decoration of the crown of Loard Krishna and also for other fancy items. It is found in wild and feral forms in most parts of the Bhartiya subcontinent.

Zoological classification

The place of pea fowl in animal kingdom is presented in the following flow chart.

Kingdom	:	Animalia
Phylum	:	Chordata
Class	:	Galliformes
Order	:	Phasianidae
Sub family	:	Phasianinae
Genus	:	Pavo
Species		1. *Pavo ristatus* (Indian Pea fowl)
		2. *Pavo muticus* (Green pea fowl)

Mutation in Indian Pea fowl (*Pavo Cristatus*)

Colour mutation has been observed in the flocks of Indian pea fowl (*Pavo ristatus*), so far there appears to be no information on further multiplication of mutants. The common mutant of peafowl is the black shoulder or Japanese mutation. In this type of mutation adult peacock is melanistic due to which wings are dark black. In another type of mutation the feathers of peacock are creamy white with brownish tipped wings. This genetic mutation produces melanin in peafowl resulting in colour dilution resulting in creamy white and brown markings. There are also occurrence of pie bald and whale colour white forms, which are considered allelie variation at particular loci. In past years white peacock were available in some of the zoos including the Lucknow Zoo.

Habitat of Pea fowl

The habitat of pea fowl spreads from semi-arid to wet forest and ranges of tropical Asia and upto an altitude of 1600 m. In some areas peafowls are found upto an altitude of 2000m. Peafowls always live near the water sources. The birds take shelter in bushes during hot sun and roost on trees, water tank or abandoned buildings in the night.

Body colour and shape of Indian Peafowl

Important measurements like body length and body weight of male and female peafowl (*Pavo cristatus*) are presented as follows.

Body length of pea cock 100-115 cm (40-45 inches) and pea hen 90-95 cm(36-38 inches). Body weight of peacock 4-6 kg (3.8- 13.2 lb) Body weight of Pea hen 2.75-4 kg (5.5 -8.8 lb).

From the measurements of the representatives of both sexes of peafowl sexual dimorphism is quite clear. The peacocks are longer, robust and much bright than the peahens. The adult peacock processes a deciduous train of long and beautiful feathers growing on feathers are 195-225 cm (78-98 inches) long. The crown of pea cock is melolic blue. The feathers on head are short and curl. The head porsesses a fan shape Cyest (Comb) made up of fine tiny feathers and carried on bare black shafts. The crest in tipped with glittering bluish green webbing. Bare white skin forms characteristic white stripes above the eyes and a white crescent patch below the eyes. The sides of head possesses glassy greenish blue feathers.

The back of pea cock is covered with scaly bronze-green feathers with blue and tan markings. The wings are buffed and barred in blue. The colour of primary feathers in buff and that of secondary feathers is black. The true tail is dark brown and covered with a seasonally deciduous train of false tail made up of long glittering feathers. True tail has only so feathers that end with elaborate eye spot except a few on either side of terminate as a crescent shaped black tip. The elongated upper tail covert is made of more than 200 long feathers. The metallic plumage colour is said to be not pigments but optical interference known as Braggs reflections that are based on regular penodic nano structures of the barbules (fibre like structures of the feathers) feathers. Slight changing in the spacing results in appearance of different colours. The brown feathers are a mixture of red and blue. One colour is produced by the periodic structure and the other one is created by a fabry-perot interference peak from reflections of two outer and inner boundries. This type of influence based structural colour is important for structural colour is important for the hues of the pea cock which change and simmer with the viewing angle, because it depends on the angle of light.

Bragg reflections- The action of a crystal in reflecting X-rays or particle wave as electrons or neutrons.

Fabry-perot interference- The effect of reflection from the inner and outer boundry of an object.

The adult pea hen also possesses a crest similar to peacock but colour is chestnut with green edged tips. The upper part of body is brownish with pale moultings. The feathers are dark brown. The lower neck is metallic green. The breast feathers are metallic greenish dark brown. The other parts of abdominal region are whitish.

The dowsing pea chicks are pale buff with a dark brown stripe on the neck connects with the eyes. There is no apparent sexual dimorphism in young peachicks. Except the wings of male pea chicks are darker (Chest nut) in colour.

Vocalization

The common calls of pea fowls are loud “may-awn” or “pian-oo” or “may-owo”. The frequency of calls increases before the monsoon season, which is also mating season for the Bhartiya pea-fowls. The vocalization also increases due to danger and emit uncommon noise for giving alert alarm to other members. A rapid “ka-ka” or “kawn-kawn” called are emitted by the pea hens for gathering, their chicks for return back to nest from foraging.

Reproduction

In pow fowls of Bharat breeding takes place during the monsoon season. The breeding in pea fowls of the Bhartiya subcontinent vary with the on set of monsoon from Sri Lanka to Himalayan Tarai. Thus, the breeding months are January to March in Sri Lanka. In the Bhartiya territory breeding season extends from June & August in north part, April to August in East part and April to May in the south parts.

The pea cocks are poly gamous. The adult males gather near the water bodies. They come together but mark their own small territory on the same ground. During breeding season adult peahens visit frequently to peacocks. A group of 5 to 7 adult pea hens can be seen near an eligible peacock during the breeding season. Exchange of peahens from one haram to another or flirting of peacock is quite common. Clowdy season stimulates the peafowls and peacocks start dancing to lure the pea hens. During dancing peacocks display the upper tail coverts spread into an arched fan and dance to attract eligible peahens for courtship. The wings are held half open and dropped. Periodical vibrations of long feathers produce gingling sound. The peacocks make different move to attract the peahens. However, peahens do not care for the luring dance movements unless they are receptive.

Before mating peahens fabricate shallow nest with breeding of soft dry grasses inside the bush for standing tall grasses. Some times they make nest in aboundened guises of buildings in isolated places.

Clutch size = 4-8 eggs of fawn to buff colour

Incubation period = 28 days

Peachicks are nidifugous and follow the mother after hatching. They are guided and trained for feeding by the mother in early life.

Feeds and feeding

The peafowl is omnivorous. Foods include grains, fruits, insects, larvae, pupae, snails, worms, small rodents and small serpents. In forests the common foods are berries, kernels, flowers, fruits and leaves besides different species of small animals. The peofowls in farming area cause great damage by eating the shown seeds in field.

This is many times fatal for the bird due to treatment of seeds with toxic chemicals before sowing. The birds near the human habitats may be seen to eat grain sorting, undegraded seeds in animals is dung, vegetables and fruits etc. In zoos they are fed compounded feeds of poultry alongwith fresh fruits and vegetables like chopped carrot, spinach, greenpea and others.

Predators

The common predators are the wild cat, feral dogs and carnivorous birds like eagle, crested hock rock eagle and owls. Peachicks are the main prey of small carnivorous birds. Although pea fowl is regarded a holy bird but sometimes people also kill peafolws particularly the pea cocks for feather and flesh. Oil of peafowl is claimed to possess mechanical value.

Uses of fallen feathers of peacock

The long feathers of pseudotail of adult peacock is collected by the people and used for the preparation of fancy and decoration articles. Fans of different designs are quite popular. In some tribal community feathers are used for the ornamentation of turbans. Ladies make neckness of various designs. They also decorate clothes with creascented part of the feather.

Green Peafowl (Pavo muticus)

It is an endangered species found east of Mayanmar and islands of indo-pacific region. This species is also crested. Wild green fowl is believed to be monogamous. The birds move in large packs of 10 to 50 birds for hunting.

Life span of Pea fowl

In the captive management of zoos Indian Peafowls have been lives for 23 year. In wild habitat life of peafowls range from few weeks for the peachicks to 9-10 years for the adults of both species. Causes of mortality are prediction, poaching and poisoning by eating sown seeds treated with toxic chemicals.

17

JANANESE QUAIL (*Coturnix cotunix*)

The Japanese quail is a common commercial species extensively bred for tender meat and egg production. The finishing period for fattening of quail chicks is very short (4 to 5 weeks). As a result turn over rate in high. The bird is also suitable for backyard farming in small and portable cages. This can be a source of additional income by feeding richen left over with little supplemental feeds. Marketing of quails 9meat and eggs) is quite easy due to high demand.

Zoological classification:

The bird belongs to class avers and order galleniferae of the animal kingdom.

Kingdom	:	Animalia
Phylum	:	Chordata
Class	:	Aves
Order	:	Galleniformi
Family		
Genus	:	Coturnix
Species	:	Coturnix coturnix

Other characteristics of quails

Body weight, adult male 100-160 g

Adult female 120-200 g

Squab at hatching 6-8 g

Females are heavier than the males.

Incubation period of eggs -14 days

Clutch size -1 to 5 eggs

Body temperature – 41.5°C or 106°F

Flour space requirement – 0.25 squar feet/bird

Sexual maturity – Male 6 weeks (40-42 days)

Female 5-6 weeks (35-42 days)

Egg production – 150 to 200 and 280 in improved strains.

Egg size – 10 g (8 to 11 g).

Egg composition – Moisture 74%

Protein 18%

Fat 11%

Others 2%

Energy value 150 -160 kcal per 100g content.

Dressing percentage – Approximately 70%

Meat quality – Highly nutritious. lEgs and breast are highly preferred.

Feed quality: Should be palatable, nutritious and balanced.

Crude protein % in feeds

0-3 weeks 26-28 %

4-6 weeks 23-25%

Layers 18-20%

ME (Kcal/kg) for growers 2800 kcal

For layers 2700 kcal

18

PARTRIDGES

The partridges are small to medium size birds represented by about 14 genera and several species. These are widely distributed in most parts of the Africa, Asia and Europe. The birds fabricate nests from dry grasses on the ground in the cover of tall and dense grasses or on the small trees in barren areas. They prefer nesting in dense bushes to keep themselves hidden from the sights of predators and poachers. These are prolific wild birds extensively captured with the use of nest and food grains. Earlier different species of partridge were available in the markets for table purpose. Now there is total ban on the hunting of wild animal species in the Bharat and many other countries. A few species of large size species are also reared as a pet for fighting competition at different festive occasions in the Bhartiya sub continent. Fighting competition between partridges is quite popular in Agra, Bareilly, Jaipur, Lucknow, Kanpur, Karanchi, Lohare, Hyderabad and many other towns. In most of the fighting games between the birds is linked with gambling.

Zoological classification of Partridge

The place of partridge in zoological kingdom is presented in the following flow chart.

Kingdom	:	Animalia
Phylum	:	Chordata
Class	:	Aves
Order	:	Galleniferae
Family	:	Phasianidae
Sub Family	:	Perdicinae
Genera	:	About 14
Species	:	Many

Some species of each of 14 genera are presented with common names in table.

Table: Genera, species and common names of some partridges.

Sl.	Genera/Species	Common names
1.	Alectris (Al) spp,	
	Al. barbera	Barberu Partridge
	Al. chuker	Chuker Partridge
	Al. graeca	Rour Partridge
	Al. magna	erzevalse Partridge
	Al. melinocephala	Arabian Partridge
	Al. philbyi	Philbyi Partridge
	Al. rufa	Red legged Partridge
2.	Ammoperdix spp. (Amm)	
	Amm. Grasioguleris	See-see partridge
	Amm. Heyi	Sand partridge
3.	Arborophila spp. (Arb.)	
	Arb. ardenis	Hainan partridge
	Arb. atroguleni	White checked partridge
	Arb. bruneopectus	Bar backed partridge
	Arb. cambochiana	Chest nut headed partridge
	Arb. charltonii	Chokalate necklaced partridge
	Arb. Chloropus	Scaly breasted partridge
	Arb. Crudigulasis	Taiwani partridge
	Arb. dpverdo	Orange-naked partridge
	Arb. gingica	White naked partridge
	Arb. hyperithra	Red breasted partridge
	Arb. javamica	Chestnut bellied partridge
	ARb. mandellii	Annalar hill partridge
	Arb. merlini	Annam hill partridge
	Arb. orientalis	Whip faced hill partridge
	Arb. rubrinosteris	Red billed partridge
	Arb. rufipectus	Sichuan partridge
	Arb. rufoglularus	Ruffus throated partridge
	Arb. sumatrana	Grey breasted hill partridge
	Arb. tarquebla	Common hill partridge

[Table Contd.

Contd. Table]

Sl.	Genera/Species	Common names
4.	Bambusicola spp.	
	Bamb. Fytchaii	Mountain bamboo partridge
	Bamb. Thoracica	China bomboo partridge
5.	Calopordox oculea	Ferrugenous partridge
6.	Heamdortgx sanguilicep	Crimson headed partridge
7.	Lerva lerva	Snow partridge
8.	Margaroper dixMedagascarensis	Medagarkar partridge
9.	Melanopordix nigra	Black blood partridge
10.	Rollulus roulroul	Creasted wood partridge
11.	Perdix spp.	
	Perd. Dauuriea	Daurian partridge
	Perd. Hodgsoniae	Thimalayan /Tiketan partridge
	Perd. Perdix	Grey partridege
12.	Ptelopachus petrosus	Stone partridge
13.	Rlogothera longirostus	Long billed partridge
14.	Xenoperdix spp.	Long billed partridge
	Xenop. Obscuratus	Rubela forest patridge
	Xenop. Udzungwersis	Udzimguva partridge

Foods of partridge

The partridges are omnivorous birds feeding on fallen grains of weeds, insects, larvae and pupae. They also eat tender leaves and flowers. Pet partridges are fed a mixture of grains, leguminous seeds, oilseeds and small fishes. Some partridge owners do not share the composition of feed mixture. Pet partridges are also allowed scavenging for an hour during dusk.

Partridge sports

The partridges are trained and groomed for fighting competition. Sponsored competitions are quite common in many town if the Bhartiya sub continent. Partridge fight competitions are common during the fares and festivals. Earlier partridge fights were also linked with gambling.

Some tales related to partridges

There are many tales and wheresay stories about the partridges in different countries.

1. In the Greek mythology the origin of partridge is believed to be the reincarnation of "Perdix", the nephew of 'Daedalus, who threw him from the top of a house due to jealous anger.
2. Pictures of partridges are used in the composition of the Christmas corals in many countries.
3. Some community of Bharat pay to caged partridges owners for libration of the bird. This is considered auspicious.

Other characteristics

Anatomy	Sexual dimorphism. Males are heavier and possess sicklas and spur
Rectal temperature	39.5 to 40°C or 103 -105°F
Respiration rate per minute	12 to 30
Clutch size	1 to 10 eggs per cycle
Annual egg production	(a) Native hen 30 to 80
	(b) Developed hen 125-300
Production types	
(a) Egg Type breeds	- White leg horn, Brown leg horn
(b) Dual purpose	- Rhode Bland Red
(C) Meat type	- Plymouth rock

Note:- Now Synthetic strains of layers and broilers are available in wide variety.

Adult body weight of non synthetic straws.

(a) White egg type fowl 1.5 to 3.0 kg

(b) Brown egg type fowl 2.0 to 4.0 kg

(c) Broiler strains 3.0 to 5.5 kg

(Now newer strains of smaller size and low fat are also available). Development of newer strains in a continuous process. There are claims for higher content of some essential amino acids.

Breeding age- 5-6 months

Breeding life- 18 to 24 months (Gutstanding hens are allowed laying for longer age upto 3 years.

Life span- Generally birds are not allowed to continue beyond 5 years. Most of them are finished at 3 years.

Photoperiod- About 14 hours light in required for optimum performance.

Daily feed requirement –

(a) Laying strains 110-130 g

(b) Broiler strains 120-150 g

Daily drinking water requirement -200 -300 ml.

19

TURKEY (*Meliagridis* sp.)

Turkey is one of the large size bird reared for fattening Demand of fattened turkey is quite high in many countries during festivals and other ceremonial occasions. Eggs are large and nutritious. Both wild and domesticated varieties are available. Fossils study of turkeys show that turkeys were present on the Earth about 23 millions year ago. At present wild species Meliagris galloparo and Meliagrus Ocellata are found in the forests of North American countries and Jucatan peninsula. Both the species were domesticated long ago in the ancient period. The domesticated version of Meliagris gallipavo became more popular. Now it is distributed widely for farming. Turkey farming is more popular in the temperate countries. Some breeds of turkey are also reared in many sub tropical countries. Small flocks can be also seen in some parts of Bharat.

Zoological classification of Turkey

The turkey belongs to class aves and order galleniferae with other details as follows.

Kingdom	:	Animalia
Phylum	:	Chordata
Class	:	Aves
Order	:	Galleniferae
Farmily	:	Meliagridinae
Genus	:	Meliagridis
Species:		1. Meliagridis gallopova
		2. Meliagridis Ocellata

Common names of some growth phases

Youngs male turkey	:	Poult, stag, Gobblor.
Youngs female turkey	:	Turkey hen.

Characteristics of Turkey

Some special characteristics of turkey has been reported as follows.

1. Some times infertile eggs of some breeds of turkey have been reported to hatch out poults. This shows the ability of parthenogenesis. There appears to be no recent record of parthenogenesis is any breed of turkey.
2. The turkeys are considered less intelligent due to a peculiar behaviour during the rains. Even during the rain storms the turkeys look up non reactive until they are washed away in the streams of rain water. This defeat has been found a genetic nervous disorder of some breeds. This condition is known as "titanic torticollar spasms".
3. Some heavy breeds of domestic turkey do not realize that they have almost lost their flying ability due to heavy body weight. Due to lack of this common sense they suffer the adverse situation of attempting flight.
4. The English navigator Williams strikcland is believed to introduce turkey in England during the sixteenth century.
5. In Europe turkey was introduced by the Spanish from the Aztacs who domesticated turkey for food (eggs and meat) and feathers. Turkey feathers were used for decoration. The humorous behaviour of mountain Turkenter god "Tezcatlipoca" liked turkey.
6. According to an English farmer Thomas Tussar the turkey were present in the farmers fare at Christarnas in 1573 A.D.
7. An aggrement took place with some farmers and navigators of England for the supply of male and female turkey to colonies in New World sent to James town in Virginia in 1607 from England.
8. At a time turkey meat was a luxury in Europe and only after 1940 AD its availability became wider due to domestic fall in price resulting from an increase in production from large scale farming.
9. In Spain, the turkey was introduced in 1500 AD.
10. While feathered turkeys are preferred because pin feathers are less visible after dressing.

Breeds of Turkey

At present several breeds of turkey have been developed in different countries. Some distinguishing characteristics of different breeds are presented in Table.

Sl.	Breed	Remarks
1.	Broad Breasted white	Most popular due to white feathers.In U.S. custom "Presidential Pordon" is received by this breed.
2.	Broad Breasted Bronze	It is second choice of the People.
3.	Standard Bronze	Triangle breasted
4.	Bourbon Red	A small size non commercial breed. Feathers are attractive dark red with white markings.
5.	Black or Spanish black or Norful black	Plumage in bright dark and attractive greenish
6.	Blue slate or slate	Feathers are grayish blue. It is a rare breed.
7.	Chacolate	It is a rare heritage breed. Its feathers have light brown markings similar to that in Spanish black.
8.	Beltoville small white	It is a small size heritage breed developed during mid twinteeth century. It is a little bigger than the midgut breed.
9.	Midgut white	It is a smaller heritage breed comparable with Betsville small white
10.	Narvagonseth	It is a popular heritage breed of turkey. This breed was developed by Naragonsett in New England.

Common by products of Turkey

Some of the common by products of turkey are utilized by the farmers and other families for the following uses.

1. Feathers are processed for the manufacture of protein supplements for the feeding of farm animals.
2. Digestive tract and other non edible offals are sterilized and fed to poultry and pigs.
3. Selected feathers are used for making fancy and decoration items.
4. Feathers are processed and blend with other fibres for manufacturing fabrics.
5. Droppings are nutritious manure for the crops, flower pots and vegetables grown in backyard garden.

Human foods from turkey

Flash is the main food for humans. Eggs of turkey are also used for the table purpose in the form of various preparations. The carcass of turkey is generally large, it is marketed in pieces. Macorated (minced) meat of turkey is also preferred for preparation of variety foods of only meat or after blending with other foods. Demand of turkey meat is high because of the very low fat content (Only 1.6). Low fat lowers the calorie, which is about 106 kcal per 100g of breast meat.

Demand of turkey meat is very high on the occasion of Christmas in many countries. Meat of white turkey is preferred. Before mid twenteenth century turkey meat was a very rare food item and it was rarely available to common families. It was a high priced luxury food.

Age of turkey at slaughter

Female turkeys are generally slaughtered younger at about 14 (12 to 16) weeks of age and toms are slaughtered at 18 (16-20) weeks of age.

Some special dishes of turkey meat

Some of the special preparation of turkey meat are leasted as follows. A few preparations have been given name for increasing palatability.

1. Pigs in blanket or rolled bacon- It is prepared from minced meat of turkey. Minced meat or liver of turkey is wrapped in bacon and then cooked to emit characteristic aroma.
2. Hot dog made of turkey meat is a special snake.
3. Roasted meat, specially breast meat and drum stick.
4. Deep fried meat in boiling edible oil.
5. Stuffed turkey- Whole eviscerated carcass is cooked after stuffing with different edible foods. Species and condiments are used for increasing flavor and taste.

Nutrients requirement of turkey (90% DM in feed)

Nutrients	NRC 1994 + other 4S values				ARC and UK			
	0-8 wk	8-16 wk	16-24 wk	Breeders	0-6 week	6-12 wk	>12 week	Breeders
1	2	3	4	5	6	7	8	9
ME (Kcal/kg)	2800	3000	3100	2850				
Crude protein (%)	28	22	16	14	30	26	18	16
Amino acids (%)								
Arginine	-	-	-	-	1.6	1.3	0.8	0.5
Glycine	-	-	-	-	0.9	-	0.7	-
Histidine	-	-	-	-	0.6	0.5	0.35	0.2
Isoleucine	-	-	-	-	1.9	1.0	0.55	0.4
Leucine	-	-	-	-	1.9	1.5	0.8	0.7
Lysine	1.60	1.25	0.90	0.80	1.7	1.3	0.8	0.75
Methionine	0.56	0.44	0.32	0.28	-	-	-	-
Meth+Cyst	1.05	0.81	0.60	0.50	1.0	0.8	0.6	0.55
Phenylalanine tyrosine	-	-	-	-	1.6	1.5	1.0	0.8
Threonine	-	-	-	-	1.0	0.9	0.55	0.4
Tryptophane	-	-	-	-	0.26	0.23	0.	
Valine	-	-	-	-	1.2	1.0	0.6	0.5
Macro minerals (% in feed)								
Calcium	1.40	1.00	0.60	2.00	0.90	1.00	0.80	1.0
Phosphorus (av)	0.70	0.60	0.50	0.60	0.45	0.50	0.40	0.5
Magnesium	0.06	0.06	0.06	0.06	0.036	0.036	0.036	0.036

[Table Contd.

Contd. Table]

Nutrients	NRC 1994 + other 4S values				ARC and UK			
	0-8 wk	8-16 wk	16-24 wk	Breeders	0-6 week	6-12 wk	>12 week	Breeders
1	2	3	4	5	6	7	8	9
Sodium	0.15	0.15	0.15	0.15	0.175	0.175	0.175	0.175
Chloride	0.15	0.15	0.15	0.15	-	-	-	-
Potassium	0.60	0.60	0.60	0.60	-	-	-	-
Micro or Trace minerals (ppm)								
Iron	80	60	60	60	96	96	96	80
Iodine	0.35	0.30	0.30	0.35	0.48	0.48	0.48	0.48
Copper	6	4	4	6	4.2	4.2	4.2	3.5
Manganese	60	60	60	60	120	120	120	100
Zinc	70	60	50	70	60	60	60	50
Selenium	0.20	01.5	0.10	0.20	0.25	0.15	0.15	-
Fat Soluble vitamins								
Vitamin A (14/kg)	11000	11000	11000	11000	12000	12000	12000	10000
Vitamin D_3 (14/kg)	1500	1500	1500	1500	1500	1500	1500	1500
Vitamin E (14/kg)	15	13	7	30	36	36	36	30
Vitamin K (14/kg)	2.5	2.0	2.0	2.0	4.8	4.8	4.8	4.0

[Table Contd.

Contd. Table]

Nutrients	NRC 1994 + other 4S values				ARC and UK			
	0-8 wk	8-16 wk	16-24 wk	Breeders	0-6 week	6-12 wk	>12 week	Breeders
1	2	3	4	5	6	7	8	9
Water soluble vitamins (mg/kg)								
Thiamin	2.0	2.0	2.0	2.0	4.8	4.8	4.8	4.0
Riboflavin	5.5	4.5	4.5	5.5	12	12	12	10
Niacin	75	65	65	45	60	60	60	50
Pantothenic acid	15	11	11	20	19.2	19.2	19.2	16
Pyridoxine	4	3.3	3.3	7	6	6	6	5
Biotin	0.26	0.22	0.22	0.26	0.12	0.12	0.12	0.10
Folic acid	1.3	0.9	0.9	1.1	2.4	2.4	2.4	2.0
Vit. B12	0.011	0.007	0.007	0.011	0.024	0.024	0.024	0.020
Choline	2000	1750	1750	1300	1760	-	-	1350
Linoleic acid (%)	1.10	0.80	0.80	1.00	-	-	-	-

Source: Nutrients requirement of Turkey; National Research council).

20

PARROTS

The parrots are common caged pet birds in the Bhartiya sub continent and many other countries. These posseses characteristic convex upper beak and ability to speak many common words and simple small names frequently used in the home. Common parrots found caged in the homes of Bhartiya families are green feathered with or without red colour necklace. Parrots are also found in other colours. Three families of parrots contain several species.

Zoological classification of parrots

The place of parrots in animal kingdom is presented in the following flow chart.

Kingdom : Animalia

Phylum : Chordata

Class : Aves

Order : Psittaciformes

Family:
1. Psittacidae (True parrots)
2. Cacatuidae (Cokatoos)
3. Strigopidae (New Zealand parrots)

Some important characteristics of true parrots are presented because these are common pet birds kept in designed cages containing a pirch inside for the perching of the bird in limited space of the cage. The family Psittacidae is a large family comprising of 5 sub families and several genera in each sub family (Table).

Table 1: Subfamilies and tribes of Psittacidae family.

Sl.	Sub family	Tribes
1.	Arinae	About 3.0 genera, 160 species and 2 lineage have been recorded.
2.	Loriinae	About 12 genera and 50 species of Lorkat and Lories groups are known. These are found in Australia, New Guinea and many islands of south pacific zone.
3.	Micropsittanae	Only 1 genera and 6 species of pygamy parrots are found
4.	Psittacinae	These are further identified in 5 min
	(i) Cyclopsitacinae, 3 genera in New Guinia.	
	(ii) Polytalini- 3 genera of broad tail parrots of Australia.	
	(iii) Psittrichadini – Single species of Peguet's parrots.	
	(iv) Psittachni – 3 genera and 12 species	
	(v) Psittaculini – About 12 genera and 70 species spread from Bharat to Australia.	
5.	Platicrecinae -	It contains 4 mini sub families, viz.
	(i) Melopsittacini-	1 species, Budgerigar only.
	(ii) Nephanini	1 genera only.
	(iii) Pezoporinini	1 genera and 2 species
	(iv) Platyceracini	8 genera and 20 species of Rozella and Relatials.

Foods of common parrots

Common parrots eat cereal grains, leguminous seeds, oilseeds and seeds of many fruits and vegetable. They of fond of fruits like guava, mangoes, custard apple and plums etc. They also eat green vegetables, tender leaves flowers and green chilli. The diets of pet parrots of vegetarian and non vegetarian households are accordingly divided. In vegetarian families parrots are fed cooked rice with pulses, milk and milk products, jaggary, green chilli, carrot, guava and other foods eaten in the family of owners. In non vegetarian families they are also fed meat, fish eggs etc.

Inteligence and learning

The intelligence and learning ability of pet parrots in company of humans have been observed as follows.

1. Parrots are intelligent avian species.
2. Many parrots can communicate with humans using simple common words used int eh family.
3. Mimicing of human speech in quite common in pet parrots.
4. A pet parrot generally call the names of children and helping hands in the home. These names are frequently called in the home.
5. Pet parrots also give informations about the visitors and strangers.
6. Some parrots understand the meaning of many common words frequently used in the home for routine works.
7. Brain, body ratio of many species of parrots in comparable with some advance species of primates.
8. Unlike other higher species the main part of intelligence in the brain of birds is the medio-rostral neostratum/ hyperstriatum ventrall. It has been claimed by the researchers that the lower part of the brain of a bird works almost similar to the brain of mankind.
9. Some parrot species are capable of using simple tools and solving some puzzles.
10. Teaching of parrots must start in early life and learning should be social learning by frequent interaction.
11. IN wild parrots learning in managed by the elders and skills are developed for playing, flight and flithts for avoiding predators.
12. Appropriate stimuli are essential for proper learning. Any neglect in routine of teaching may result in retardation of learning and retention of words in the mind. Therefore, repeated practice is very important.

Merits and demerits of parrots

1. Pet parrot (s) in cage is a good comparison for the children, elders, lonely person in the home and pet dog(s). Cat is enemy number one because caged parrot is an easy prey.

2. Pet parrot also works like a watchman and starts characteristic calls on sight of any stranger in the premices of house.

3. It is a source of solace for the lonely persons idle in the home.

4. Wild parrots live in large groups. These are pest on fruits and crops. Parrots not only eat the fruits and pods but also cut and destroy unripe fruits and immature crops.

Life span of parrots

Long life has been recorded for the parrots of many species. Common Bhartiya pet parrots maintained properly in cages often exceeded half century. Higher life span exceeding 80 years have been recorded in some of the species belonging to cockatoos, macaws and Amazon parrots.

21

PIGEON (*Columba livea*)

The pigeon is a cosmopolitan bird found in all the continents from tropics to temperate zones. Pigeons, at least a pair of pigeons are seen by the pilgrims visiting highly pois cave of Lard Shiva, the Amarnath situated in the Jammu and Kashmir state of Bharat. This lovely bird was initially domesticated for fun and sports. Later on it was also recognized as a food bird. There are several stories of the services of pigeons as a messenger or postman. Such tales are quite common in most of the Arabian countries and Indian sub-continent. The clutch size of pigeon is through small but fertility is quite high due to when they multiply at a fast rate. Pigeon is a docile bird and its hunting for meat is considered a cowar dish act in most parts of the Bharat.

The wild grey pigeon flock prefer abandoned buildings for living in the holes and windows. They prefer safe sites for laying to protect eggs and squlabs from the predators. Common predators are owl, eagle, hocks and reptiles.

At present there are several breeds and varieties have been developed in many countries. Different breeds and varieties of pigeon have been evolved by breeding and selection for show, sports, recreation and also for table purpose. Pigeon breeding for meat production is quite common in Assam and sister states of the Bharat. Pigeon squabs for meat are marketed at 5-6 weeks of age and 150-200 g body weight. Heavy breeds of pigeon for fattening have been developed in France and other countries, which grown upto 600 g or more in 5-6 weeks. Pigeon meat is a delicacy in many countries. High content of cholesterol has been claimed in pigeon meat, when compared with the chicken meat.

Zoological classification of pigeon

The pigeon belongs to class eves and family columbiformiae as presented in following form.

Kingdom	:	Animalia
Phylum	:	Chordat
Class	:	Aves
Order	:	Columbiformes
Family:	:	Columbinae
Genus	:	Columba
Species	:	Columba livea

Foods of pigeon

Wild pigeons eat fallen grains of weeds, cultivated crops, herbs and shrubs. They also eat tender leaves. Domesticated pigeons are generally fed the compounded feeds of chicken and broilers.

Body weight

Body weight of adult pigeons of different breeds ranges from 200 g to 800 g. Female gain more weight before breeding and laying. There is no distinct sexual dimorphism in the body weight. However, many pigeon breeders claims to identify the sex of pigeon by visual appraisal.

Plumage colour

Colour of plumage is highly variable in the domesticated breeds. Plumage colour adds to beauty of the pigeons. The plumage can be single colour or combination of two or more colours in different shades. Some of the common plumage colours are grey, brown, white, golden, bluish brown and radish white etc.

Weight of eggs- 6 to 12g

Incubation period-1 7 (16-18) days

Weight of squab at hatching- 20 50 g.

This depends on the body size of breed

Feeding of squab- Pigeons are different than most of the other avian species, because they secret crop milk for the feeding of newly hatched squabs in early life. The crop milk is highly nutritious. It contains 11 to 19 percent protein, which is rich in free amino acids. Free amino acids constitute 15-20% of the total nitrogenous constituents. It is also a rich source of fat which constitutes 4.5 to 12.7%. Fat of crop milk is composed of 80% triglycerides, 12% phosphor lipids and remaining 8% are other lipids and lipid soluble constituents like fat soluble vitamins and others. A growth factor is also considered to be present int eh crop milk.

Some physiological values of pigeon

Some of the physiological values of the pigeon are as follows.

1. Body temperature, 106-107oF or 41-42oC.
2. Respiration rate, 20-35 per minute.
3. Clutch size – 2-5
4. Floor space requirement- 20 cm x 25 cm
5. Feed intake per day – 100-200 g in different size of adult pigeons.
6. Sexual maturity -20 to 25 weeks in bothsexes.
7. Breeding life-About 10 years in both sexes.
8. Age at slaughter for meat – 8 (6-10) weeks.
9. Sports – Different styles of flight. Some pigeons lure members of other group to their group during flight.

Classification of pigeons in families

The methods of grouping pigeons in different families is highly variable between the countries and nitrogen within the country. One of the method has following 7 families, i.e. (1) colour, (2) owe, (3) puter, (4) squabbling (5) sporting, (6) structure and (7) voice.

Breeds of pigeon

Some important breeds of pigeon alongwith salient features and other notable traits are presented in table.

Table : Some popular breeds of pigeon.

Sl. No.	Breed	Characteristics
1.	Archangel/ Gimpee	A German breed of toy group, briantally coloured and attractive. Body is small, slender and elegent.
2	Carriers	These are descendents of the old homing pigeons of Persia evolved for carrying messages. It was mentioned in the book of Charle's Darwin-entitled variation of Animals and plants on Domestication.
3	Damascene/ Mahonets	Origin of this breed is not known. The pigeon has been carved on Egyptian structures and presented in literatures. It is believed that probably Damuscus in Syria is the place of origin. Other group believe Iran or Turkey.
4	Dutch Capuchin	The origin of this breed is not the Netherlands. The breed was going to be extinct during the second world war, when some Dutch people brought, bred and developed in the present form.
5	Drummer or Trumpeter	This is a widely distributed pigeon breed due to its characteristic vocalization. This breed is known as Drummer in Germany whereas in America and England it is called Trumpeter. It has entered in Europe from Bukhara via central Asia and then carried to America. The breed is popular in Russia and called Russian Trumpeter.
6	Egyptian Swift	It is a composite breed of at least 10 distinct families of varieties. These families were conserved for the maintenance of distinct identity of the breed.
7	Fan Tail	This is an Indian breed which has been separated in several new breeds in different countries. The original breed is now known as Indian Fan Tail. Most breeds of Fan Tail were evolved in England and Scotland.
8	Foot Soldier/Flying Serpent	These two names of the breed have been derived from the elegent gait and red eyes. The voice of this breed is roaring through a gasping mouth.

[Table Contd.

Contd. Table]

Sl. No.	Breed	Characteristics
9	Helmet	This a classic ancient breed of pigeon constituted of several varieties of different plumage colour, markings on the tail and cap or helmet. Beak and crest also differ among the varieties.
10	Jacobin	The origin of this breed is claimed to be India and Cyprus by different groups. This breed is popular in England as a bird of aristocracy.
11	Komarner Tumbler	The origin of this breed is Germany and now liked in many countires. The modern Komorner of different countries may be distinguished by difference in the colour of plumage and some other features.
12	Modena	This breed was developed in Modena town of Italy during thirteenth century. This is a famous fighting breed. The breed spread to England, Germany and other countries.
13	Nun	The Nun is a medium size breed of cobby and well balanced body. The pigeon is full of action and posseses well developed protruding breast, short legs and small feet. The crown, bib, primary feathers and tail are coloured, while remaining part of the body in bright white. The crest is smooth and white.
14	Old German owl	This attractive breed was evolved in Germany and has gained popularity in the united states. The breed in friendly and its breeding is easy. This breed does not need external assistance for rearing its squab.

Feeding of domestic pigeon

The pigeons are reared for recreation, companion and gambling. The other groups are fatt3ened for the production of delicious tender meat. Meat breeds of pigeons are popular in many European countries and also in north-east sister states of Bharat. Compounded feeds for broilers of quails are fed to fattening pigeons. Other group of pigeons are allowed daily flight and outing. During this period they also feed insects, worms and larvae etc. In the yard they are fed a mixture of cereal grains, pulses and oil seeds. Some owners also offer small fishes (Fresh or dried).

The diets of free living pigeons include variety of fallen grains of crops and weeds, larvae, pupae, worms and tender leaves. They also engulf small grits of stone or brick.

22

BULBUL (*Pycnonotus*)

The bulbuls are found in the tropical forest and orchards of the juicy and paepy fruits. The bird is more common in the Bhartiya sub continent, south-east Asia, Africa, islands of south–east pacific region, Arabian countries, Australia, New Zealand, Fizi, Tonga and Hawai islands. Two varieties of known bulbuls are the (1) red vented bulbul and (2) yellow vented bulbul. The preferred places of nesting and perching are the fruit trees. Pet bulbul is an ornamental as well as bird for games. Like few other caged birds bulbuls are also trained for fighting competition often linked with gambling. In the Bhartiya sub-continent famous cities of bulbul fight sports are Lucknow, Varanasi, Agra, patna, Jaipur, Hydarabad, Lohare, Karanchi and other places.

Zoological classification

The place of bulbul in animal kingdom is presented in the following flow chart.

Kingdom	:	Animalia
Phylum	:	Chordata
Class	:	Aves
Order	:	Passeriformes
Family	:	Pyenonotidae
Genus	:	Pyenonotus
Species	:	1. Pycnonotus Cafer
		2. Pycnonotus goiavier

The red vented bulbul (Pycnonotus cafer) has been studied more extensively than the yellow vented bulbul (Pyenonotus goiavier). The main apparent difference between the two species is the colour of vent area, which red in red vented bulbul and yellow in yellow vented bulbul. The other difference is the crested head of red vented bulbul. The body is dark brown and head is blackish. Posterior abdominal part is much lighter grayish in colour but vent is bright red.

Sub species of bulbuls

Several sub species of red-vented bulbul have been identified (Table1).

Table 1: Sub species of bulbuls.

Sl. No.	Sub species	Dominant habitats
Picnonotus cafer = Red Vented bulbul		
1.	P.C. Intermeditus	West part of Kashmir
2.	P.C. Pygenus	Lower Himalayan region
3.	P.C. bengalensis	Himalayan Nepal to North east states of Bharat.
4.	P.C. pallidus	Tropical zone of Bharat
5.	P.C. Saturatus	Godawari delta region
6.	P.C. hunayuni	Lower Sindh region
7.	P.C. Stand fordii	-do-
8.	P.C. melanchimus	South Myanmar
9.	P.C. burmanicus	N.E. Bharat & north Myanmar
10.	P.C. nigropilius	Southern Myanmar
11.	P.C. Chryscorrhoides	China
12.	Picnonotus gravier	Yellow vented bulbul Bhartiya sub continent, Indonesia, Philippines and adjoining islands.

Reproduction

The breeding season of red vented bulbul extends from June to September. Clutch size is 2-3 eggs. Eggs are small and dull yellow pinkish with red spots. Several clutches are laid in a season. Before mating both parents fabricate together nest for the incubation of eggs and housing of chicks before outing. Incubation period is 14 days Both parents feed the squab in early life upto the growth of feathers for flight. Squabs are guided for flight and food identification.

The clutch size of yellow vented bulbul is 2-5 eggs and thus species is also multi clutch layer in a breeding season.

Feeds and feeding

The feeds of bulbul include grains seeds, oilseeds, fruits, flowers, leaves, insects, larvae and pupae etc. The bulbuls are unable to synthesize vitamin C and requires dietary sources. Bulbuls also sip nectar of flowers. Pet bulbuls are fed guava and other juicy fruits for the supply of vitamin C.

Some cultural significance of bulbul:

A few socio-cultural significance of bulbul has been listed as follows-

1. Bulbul fighting sport was quite popular during the monetinth century in many parts of Bhartiya sub continent.
2. Some people tame a bulbul and carry on fist or palm in the market and locality. However, tamed bulbuls are kept secured by tieing/ tethering a string in one of the legs.
3. A metallic or wooden perch is provided to bird during show at any gathering.
4. Expansive necklace and other jwelleries are often put on the bird by rich owners.
5. Bulbul fight is also popular in the towns of cock fight games.

Common name of bulbul in different parts of Bhartiya sub continent:

The common names used for the identification of bulbul in different parts of Bhartiya sub continent are listed as follows.

1. Hindi = Bulbul, Kala bulbul, Guldam
2. Assansese = Bulbuli soai
3. Nagami = Insi bulip
4. Lepcha = Manclaph
5. Oriya = Bulubul
6. Marathi = Lalbudya bulbul
7. Himachali = Kala Painju
8. Gujrati = Hadiya bulbul

9. Malyalam	=	Naltu Bulbul
10. Cachhari	=	Dao bulip
11. Kannada	=	Kempn dwarda pikalora
12. Bilpart of Gujrat	=	Peetrolys
13. Tamil	=	Kondalati, Kondi-kurvi
14. Telugu	=	Righi-pita
15. Bhutani	=	Paklam
16. Sinhali	=	Konde Kumlla

23

MAYANA/MYNAH/MYNA (*Strunidae Birds*)

The common name Myana or Maina was given to this starling bird in the Bharat, which subsequently spread universally. Myana is not a single species but a group of person birds comprising of many genera and species originally found in the Bhartiya sub continent and other regions of south and east Asia. The birds have been spread widely and now available in almost all countries and continents. Mayana, specically the Bhartiya talking Mayana capable of mimic human and other voices when caged as pet in houses. These are gregarious and live in groups of several families in the wild state.

Zoological classification of Mayana/ Myana

The bird comprising of about 10 genera has been placed as follows in the animal kingdom. Red vented bulbul has 11 known varieties.

Kingdom	:	Animalia
Phylum	:	Chordata
Class	:	Aves
Order	:	Passeriformes
Family	:	Sturnidae
Genera	:	1. Acridothesis
		2. Ampaliceps
		3. Basilornis
		4. Enodes
		5. Gracula
		6. Leucopsar
		7. Mino
		8. Scissirostrum
		9. Streptocitta
		10. Sturnus

Table: Species are given with common name.

Sl. No.	Species	Comman name
1.	Acridothesis otocinatus	Collered Mynah
2.	Acro. Bermanecus	Vimis breasted Myana
3.	Acro. Cinerus Pale bellied Mynah	
4.	Acro. Cristatollus Crested Mayna	Himalayan Nepal to North east states of Bharat.
5.	Acro. Fuscus	Jungle myana
6.	Acro. Ginginianus	Bank Myana
7.	Acro. Grandis	White vented Myana
8.	Acro. Javanicus	Janani Myana
9.	Acro. Tristis	Common Myana
10.	Ampeliceps Oronatus	Golden creasted Myana
11.	Basilornis celevensis	Sulenesi Myana
12.	Basil. Corytheise	Long Creasted Myana
13.	Basi. Galiatus	Helmeted Myana
14.	Brassl. Miranda	Apo Myana
15.	Enodes erythrophris	Fiery bround Myana
16.	Gracula enganensis	Southern hill Myana
17.	Grac. Indica	Sri Lankan Myana
18.	Grac. Ptelogenys	
19.	Grac. Religiosa	Common hill Myana
20.	Grac. Robusta	Nias Myana
21.	Lencopsar rothochildi	Bali Myana
22.	Mino anais	Golden
23.	Mino dumontii	Yellow faced Myana
24.	Mino Kreffti	Long tailed Myana
25.	Scissirostrum dubium	Finch bellied Myana
26.	Streptocitta albertinae	Bared-eyed Myana
27.	Strep. Alticollus	White neck Myana
28.	Sturnus cinar aceus	White chick starling Myana
29.	Stu. Contra	Asian pied sterling Myana
30.	Stu. Melanopterus	Black wing sterling
31.	Stu. Cericeus	Red billed sterling

Habitat and behaviour

The Myanas (Mynalus) are medium size passerines birds, which are found on the outer trees of forests, orchards of guava, mango and other drupe type fruits. They also make nest on scattered trees on the grasslands. Myana prefers to live in the cavities of trees. They mimic human voice and copy small words.

The myanas are territorial and aggressive. Many species of myana prefer to live near the human colonies near the forest and orchards.

Morphology of Myanas

The morphology of myanas is quite variable. Plumage is dark brown to black in most of the species. There are characteristic markings and distinguishing colour of the bill, vent, breast or other parts. Body size is also variable among the species.

Foods of myana

Like most of the avian species the myanas (Mainas) are also omnivorous. The diets include food grains, seeds, insects, larvae, pupae, worms and small snails. They like to eat juicy frits and tender leaves. Some times they also eat small reptiles.

Reproduction

The reproductive behaviour of myanas is considerably variable among the species. In most of the species courtship and pairing begins before the breeding season. Pairs of myanas make nest in the cavities of trees in advance. Some times they also fabricate nest on the isolated projections and sheltered parts on outer parts of a building. Such places for breeding are generally preferred by the species living near the human habitations of rural areas.

Clutch size = 3 to 5 eggs.

Incubation period = 11 to 14 days in different species.

Incubation is carried by both parents.

Squab (chicks) are nourished by both parents until wings are well developed for flying.

The clutch size of yellow vented bulbul is 2-5 eggs and thus species is also multi clutch layer in a breeding season.

Feeds and feeding

The feeds of bulbul include grains seeds, oilseeds, fruits, flowers, leaves, insects, larvae and pupae etc. The bulbuls are unable to synthesize vitamin C and requires dietary sources. Bulbuls also sip nectar of flowers. Pet bulbuls are fed guava and other juicy fruits for the supply of vitamin C.

Some cultural significance of bulbul

A few socio-cultural significance of bulbul has been listed as follows-

1. Bulbul fighting sport was quite popular during the monetinth century in many parts of Bhartiya sub continent.
2. Some people tame a bulbul and carry on fist or palm in the market and locality. However, tamed bulbuls are kept secured by tieing a string in one of the legs.
3. A metallic or wooden perch is provided to bird during show at any gathering.
4. Expansive necklace and other jwelleries are often put on the bird by rich owners.
5. Bulbul fight is also popular in the towns of cock fight games.

Common name of bulbul in different parts of Bhartiya sub continent

The common names used for the identification of bulbul in different parts of Bhartiya sub continent are listed as follows.

1. Hindi = Bulbul, Kala bulbul, Guldam
2. Assansese = Bulbuli soai
3. Nagami = Insi bulip
4. Lepcha = Manclaph
5. Oriya = Bulubul
6. Marathi = Lalbudya bulbul

7. Himachali = Kala Painju
8. Gujrati = Hadiya bulbul
9. Malyalam = Naltu Bulbul
10. Cachhari = Dao bulip
11. Kannada = Kempn dwarda pikalora
12. Bilpart of Gujrat = Peetrolys
13. Tamil = Kondalati, Kondi-kurvi
14. Telugu = Righi-pita
15. Bhutani = Paklam
16. Sinhali = Konde Kumlla

24

SARAS CRANE (*The Grus*)

These Saras is a large size tall bird of tropical plains commonly seen on the waste/ fallow lands in the cultivated areas of the Bhartiya sub continent. These are non –migrating birds prefer to live on the wet lands and near the water sources. It is also found in the north-east parts of Asia and Australia. The Saras is a tallest living flight bird, which pulled back its legs during flight. This species is conspicuously different from the other cranes for its red head and upper part of the neck on a grey feathered body. The bird is known for fidality.

Zoological classification of Saras crane

The place of Saras among the aves of the animal kingdom is presented in the following flow chart.

Kingdom : Animalia
Phylum : Chordata
Class : Aves
Order : Gruiformae
Family : Gruidae
Genus : Grus
Species : Grus antigone

Sub species with common names are as follows

	Sub Species	Common name
1.	Grus antigone gilliae	Australian saras
2.	G.a. antigon	Janghil (Indian Saras)
3.	G.a. sharpie (Sharpii)	Bermese, Indo-china
4.	G.a. luzonica	Philippines (extinct)

Morphology

1. It is a largest living flying bird, which is also non-migratory.
2. The height of adult Saras reaches upto 180 cm (6 feet) in a range of 115 to 180 cm (46 to 71 inches).
3. The females are shorter than the males but there is no visible sexual dimorphism.
4. The legs are very long and pink.
5. The head and continuing upper part of the neck is bare and red. The red colour becomes bright during the breeding season.
6. Body weight ranges from 5 to 12 kg (11 to 25 lb) and sharpie may be heavier.
7. Body feathers are grey.
8. Wing span extends from 200 to 250 cm (88 to 100 inches).
9. The Australian Saras is smaller than the Bhartiya saras.
10. The hind toe is placed high and reduced in size.
11. The Bhartiya Saras contains a white collar below the bare red head and white tertiary rameges.
12. The plumage of Australian Saras is darker and a larger grey patch of ear coverts is also present.

Behaviour of Saras

1. The Saras Crane is normally non-migratory. Short distance migration occurs in search of water sources.
2. South-east Asian population of G.a. sherpei is migratory.
3. Breeding pairs mark their territory and defend it from the intruders by vocal calls and chasing.
4. Non breeding birds mostly include the juveniles and live in flocks of different size ranging from few to hundreds. Solitary birds are also seen in the fields near villages. Large size flocks of 100 to more than 400 members are seen on wet lands and lakes of Madhya Pradesh, Uttar Pradesh, Gujarat, Rajesthan and other states of Bharat.

5. Non-breeding birds are expelled from the territory of breeding pairs during the breeding season. Probably such individuals lead a solitary life in isolation.
6. Moulting in adults occurs after 2-3 years.

Feeds and feeding.

The Saras is omnivorous bird. They eat grains, seeds, insects, larvae, pupae, snails, frogs, small fishes, aquatic herbages and roots. Saras can be seen killing and eating small snakes (Xenochrophis piscator). They also eat eggs of other birds of the locality.

Reproduction

1. The Saras breeds round the year but monsoon season is most preferred. Intensive breeding occurs from July to October in most parts of the Bhartiya sub continent and wet season in Australia.
2. Pairing of the Bhartiya saras is life long and death by starvation has been reported on the death of partner.
3. Nest is fabricated with reed and dry grasses in shallow water or on the wet land on an elevated place. The size of nest is about 1 metre x 1 metre (about 10 square feet). Old nests, if available, are also repaired and used for laying and incubation of the eggs. A pair often use their nest for 4-5 times.
4. Clutch size –Normally 1 to 2 , rarely 3 eggs.

   Egg colour –dull grey

   Egg weight – 200 to 270 g (mean 2408g).

   Incubation period -27 -35 (31) days.

   Egg shell on hatching is separated and eaten by the parent.
5. On suspecting danger they cover eggs with dry grasses.
6. Parents fed the chicks for first few days and then chicks start to follow the parents to learn foraging.
7. The chicks are trained to freeze and hide in grasses for protection from predators. For this parents make voice as "karr-rr-rr".
8. Growing chicks live and move with parents for about 3 months.

9. Incidence of divorce and elopement is rare in the Bhartiya Saras.
10. About 42 years life of Saras has been recorded in Zoo.

Enemies of Saras

1. Wild and common crow (Corvus spp.) damage eggs in nest.
2. In Australia, wild dog (Dingo) is predator of the bird.
3. Kites particularly the Brahimini kite (Halisastur indicus) eat eggs of Saras.

Saras in mythology and culture

1. The first poet on the earth "Maharishi Valmiki composed first ever known verse on seeing the crying mate on killing of partner by a fowler during the courtship. This instance is more than 5000 years old (The time of Ramayan epic).
2. Saras is a state bird of Uttar Pradesh and protected by the government.
3. The strong fidality in Saras pair is highly esteemed and in the Gujarat state newly wedded couple are taken out in fields to see a pair of Saras.
4. Mughal emperor Jahangir (1607 AD) observed that Saras pair in Gangatic plains always laid only two eggs at an interval of 48hours and incubation period was 34 days.
5. Eggs of Saras are used in local medicines by some tribal groups.
6. Successful breeding in captivity was recorded in 17 th century by the Emperor Jahangir in Bharat.
7. Young chicks can be reared by hand feeding and tamed like a pet dog.

25

SWAN (*Cygnus* spp)

The Swan is an attractive beautiful aquatic bird of cold climate. Most species live in temperate countries and annually migrate to tropical water bodies during the winter season. The swan has important place in the mythology and ancient literatures of the many countries specially that of Bhartiya sub continent, other Asian and European countires. Due to monogamous relationship the swan is a symbol of true love. The white species of swan is considered a bird of heaven, which is the ride of the Goddess Saraswati, the goddess of learning and wisdom in the Bhartiya sub continent. White swan live in the lakes of isolated places in the Himanaya.

Zoological classification of Swan

The place of swan (Cygnus spp.) in animal kingdom is presented in the following flow chart.

Kingdom	:	Animalia
Phylum	:	Chordata
Order	:	Anseriformes
Family	:	Anatidae
Sub family	:	Anserinae
Genus	:	Cygnus
Species	:	Many presented in table

Some of the species of swan (Cygnus) of old world and new world are presented in the following table.

Table: Species of Swan (Cygnus spp.)

Species	Common name	Location
Cygnus atratus	Blue swan	Australia, New Zealand
C. buccinators	Trumpeter swan	North America
C. columbianus	Tundra swan	Tundra zone
C. bewikii	Bewiek's swan	Asia, Europe
C. Cygnus	Whooper swan	Asia, Europe
C. melan coryphus	Black naked swan	South America

Fossils of swan have been also found in some excavations.

Some special characteristics

Some special characteristics of swan different than most of the other avian species are presented as follows.

1. Some varieties of swan like whooper, mute and trumpeter are the largest flying birds.
2. Most of the species of swan are migratory. They normally move from temperate to tropeical water bodies during the winter season.
3. Pure white specimens are found in many species of north hemisphere. The swans of south hemisphere are mostly blue and white and those of Australia are black except the white flight feathers on the wings.
4. The swan is an extensively referred bird in the mythology and folk lores of many countries including the Bhartiya sub continent.
5. Fossils show that many species of swan are extinct.

General morphology of swan

The largest variety of swan is about 150 cm (60 inches) long and 15 kg (33 lb) in weight. The wing span may be upto 3 meters (10 feet). The neck and feet are long. In the adults skin between the eyes and base of bill is featherless. The long bones are pneumatic due to which swans fly high and longer. The legs are blackish grey in most of the species and pink in two species of south America. The long neck, elongated spindle shaped body and broad webbed feet are highly useful for swimming and floating on the water. The colour of bill is blue, yellow, red and black in different species. The bill and neck are evolved to catch fishes and other aquatic edible creatures.

Foods of swans

In the Bhartiya mythology the swan is believed to eat pearl and wholesome milk and they are considered capable of separating added water in the milk. As per belief of eating pearl probably due to observation of eating the pearl producing mussels. In fact swans are omnivorous birds preferably feeding on fishes, other aquatic animals and plants. They also eat grains, roots, tubers and leaves. The meal of diet is made of different herbaceous foods including level, lemna and azolla etc.

Reproduction in swan

Earlier swans were believed to be monogamous couple but later studies revealed that divorces as well as adultery are not uncommon in the species. However, in most of cases poor bond last long and continue even during migration as evident in the pairs of Tundra swan.

Egg weight = about 340 g

Egg measurements = 11.3 cm x 7.4 cm

Incubation period = 34 to 45 days

Male mate of the pair helps in the fabrication of nest but incubation is done by the matter in most of the species studies so far.

Swan in mythology and culture

In the Bhartiya mythology and culture the swan, particularly the white swan is considered a bird of heaven.

1. White swan is the ride of Godess Saraswati, the diety of learning in sanatan system.
2. Swan (Hansa) is a highly revered bird famous for piousness and judgement, capable of separating added water from milk as said in the following vepe.

   " Hansa swetah, buckah swetah" kah bhedah hansa bakayo? Near-Ksheera viveketu, hansa hansa, buckah=buckah".

The meaning is both Hansa (swan) and bakah (Duck) are white. The ability of swan to separate milk from a mixture with water identify swan from the duck.

3. The summer habitat of swan is the lake " Mansarover" on the high Himalaya and winter is spent in the lake of Indian plains.
4. The revered Saint detached them from the worldly things are respected as "Paramhansa" as they are like feathers of swan that do not wet even on diving in water.
5. Meharishi "Kashyap" named the eight daughters of king "Dakschha" and one of them gave birth to swan.
6. Shishupal insulted Bhisma and forecasted before death Bhisma will see the fate of the birds that died in search of food and their eggs were eaten by the swan.
7. King Nal patromised a swan to serve as messenger.
8. Swan is considered a symbol of justice in Bhartiya sub continent.
9. "The myanas are territorial and aggressive. Many species of myana prefer to live near the human colonies near the forest and orchards.

SUBJECT INDEX

C

G

H

I

J

K

L

M

N

O

P

Q

R

S

Printed in the United States
by Baker & Taylor Publisher Services